PRELIMINARY FIELD REPORT

OF THE

UNITED STATES GEOLOGICAL SURVEY

OF

COLORADO AND NEW MEXICO,

CONDUCTED

UNDER THE AUTHORITY OF HON. J. D. COX, SECRETARY OF THE INTERIOR,

BY

F. V. HAYDEN,
UNITED STATES GEOLOGIST.

WASHINGTON:
GOVERNMENT PRINTING OFFICE.
1869.

LETTER TO THE SECRETARY.

DENVER, COLORADO TERRITORY,
October 15, 1869.

SIR: In accordance with your instructions dated Washington, April 1, 1869, I have the honor to transmit my preliminary field report of the United States geological survey of Colorado and New Mexico, conducted by me, under your direction, during the past season. A portion of your instructions is as follows:

"You will proceed to the field of your labors as soon as the necessary arrangements can be made and the season will permit, and your attention will be especially directed to the geological, mineralogical and agricultural resources of the Territories herein designated; you will be required to ascertain the age, order of succession, relative position, dip, and comparative thickness of the different strata and geological formations, and examine with care all the beds, veins, and other deposits, of ores, coals, clays, marls, peat, and other mineral substances, as well as the fossil remains of the different formations; and you will also make full collections in geology, mineralogy, and paleontology, to illustrate your notes taken in the field."

In accordance with the above instructions I proceeded to Cheyenne, Wyoming Territory, where my preparations and outfit were made.

My assistants were selected as follows:

1. James Stevenson, managing director and general assistant.
2. Henry W. Elliott, artist.
3. Rev. Cyrus Thomas, entomologist and botanist.
4. Persifer Frazer, jr., mining engineer and metallurgist.
5. E. C. Carrington, jr., zoologist.
6. B. H. Cheever, jr., general assistant.

Five men were also employed, three of them as teamsters, one as laborer, and the other one as cook.

As soon as my preparations were completed, my field labors commenced, June 29, at Cheyenne. Limited somewhat as to time and means, I arranged my plans so as to cover as much ground as possible and secure the greatest amount of geological information. On the plains the geological structure is very simple, and frequently over large areas the basis rocks are concealed by superficial deposits. It seemed best, therefore, to make my examinations southward along the eastern base of the Rocky Mountains for the purpose of studying the upheaved ridges, or "hog backs," as they are called in this country. These ridges afford peculiar facilities for working out the geological structure of the country. Indeed, they are like the pages of an open book upon which the geologist can read what the Creator has written upon each formation known in the country from the granite mass that forms the nucleus of the loftiest mountain range to the most recent tertiary formation inclusive. Often in a little belt, from half a mile to four or five in width, one may travel over the upturned edges of nearly all the formations in the geological scale, and the opportunity was presented, in this way, for tracing out their relations by studying the junction of the changed with the unchanged rocks.

From Cheyenne to Denver we examined with some care the mines about the sources of the Cache à la Poudre River and the coal mines at South Boulder. From Denver we visited the silver mines at Georgetown, and the gold mines of Central City, thence to the Middle Park, where we found much of interest geologically. We then returned to Denver and pursued our way southward, passed the "divide" to Colorado City, Soda Springs, Cañon City, Spanish Peaks, Raton Hills, Fort Union, Mora Valley, Santa Fé, Placiere Mountains, &c. Along this route the scenery was grand beyond description. At Colorado City there is an area of about ten miles square that contains more material of geological interest than any other area of equal extent that I have seen in the west.

The coal formation along the base of the mountains was studied with great interest. With these coal beds are associated valuable deposits of brown iron ore. The coal and iron deposits of the Raton Hills extend from the Spanish Peaks to Maxwell's, and the supply of both is quite inexhaustible and of excellent quality. The future influence of these two important minerals at this locality, on the success of a Pacific railroad, cannot be over-estimated. It is believed that the coal and iron mines of the Raton Hills will be of far more value to the country than all the mines of precious metals in that district.

The next locality for coal was at the Placiere Mountains. In one locality here, the coal has been changed into anthracite by the eruption of a basaltic dike, the igneous material of which had poured over the coal strata. Vast quantities of brown iron ore are associated with this coal, and magnetic iron ore is found in the gneissoid rocks of the mountain. The gold mines here are very rich and are now wrought upon a true scientific plan.

From Santa Fé we proceeded up the Rio Grande through the San Luis Valley, Poncho Pass, Arkansas Valley, through the South Park to Denver again. We could only give a glance at the salt springs and gold mines of the South Park, but we gathered much valuable information in regard to this interesting region. To the geologist Colorado is almost encyclopedic in its character, containing within its borders nearly every variety of geological formation. The portion of the country examined by me this summer, comprises a belt about five hundred and fifty miles in length from north to south, and almost two hundred in width from east to west.

The collections in all departments are very extensive and valuable, comprising geological specimens, fossils, minerals, plants, birds, quadrupeds, reptiles, and insects, all of which are to be arranged and classified in the museum of the Smithsonian Institution according to a law of Congress.

My report, herewith transmitted, has been written under circumstances of great pressure at odd moments, in traveling from point to point, or in camp after the labors of the day were completed, far away from books or any opportunities for careful elaboration. It may therefore be regarded as little more than a transcript of my field-notes.

Accompanying my own report will be found those of my assistants. Mr. Persifer Frazer, jr., on the mining resources of the route passed over, and Mr. Cyrus Thomas on the agricultural resources. I regard these reports as of great practical value to the country.

I take this opportunity of tendering my thanks to all of my assistants for their cordial co-operation throughout the entire survey. The reports of Messrs. Thomas and Frazer will speak for themselves. Mr. Elliott, the artist, has labored with untiring zeal, and has made more than four hundred outlines of sketches, and about seventy finished ones for the final reports. Each one of these sketches illustrates some thought or principle in geology, and, if properly engraved, will be invaluable. My principal assistant, Mr. James Stevenson, who has been associated with me in my western explorations for many years, has rendered me indispensable services throughout the entire trip.

I beg permission to state here that my appropriation was so limited that had it not been for the kindness and generosity of the military authorities of the country, I could have accomplished but a small portion of the work that I have performed during the present season, and I take this opportunity to say that the West is very largely indebted to them for whatever benefit my labors have been or may be to the country.

Before leaving Washington, I made application by letter to General Sherman, commanding the armies of the United States, for such assistance from the military authorities of the West as could be afforded to me without manifest injury to the public service. On my letter of application, General Sherman placed the following indorsement:

"This application is referred to the commanding officers of the departments, districts, and posts, who will extend to Professor Hayden's party the usual courtesies, and the privilege of purchasing a limited quantity of provisions on the same terms as officers."

Similar indorsements were made by Generals Sheridan, Schofield, and Augur. The greater part of my outfit was supplied to me by Colonel E. B. Carling, United States army, depot quartermaster at Fort D. A. Russell, Wyoming Territory; and I cannot express too cordially my grateful acknowledgments to him for his generous aid, not only for this season, but also for two previous campaigns. I am also under equal obligation to General William Myers, United States Army, chief quartermaster department of the Platte, at Omaha, for invaluable aid in several past years. When we came in the vicinity of a military post, at Fort Union, Santa Fé, or Fort Garland, we received all the aid we needed.

I would also extend my grateful acknowledgments to the press and the citizens of Colorado and New Mexico, but more especially to Colorado for their cordial aid and sympathy in all my explorations.

If my labors have added anything to the sum of human knowledge and the honor of our country, I shall be content.

I remain very respectfully, your obedient servant,

F. V. HAYDEN,
United States Geologist.

Hon. J. D. Cox,
Secretary of the Interior.

REPORT OF F. V. HAYDEN

GEOLOGICAL REPORT.

INTRODUCTION.

In order that the relation of the different geological formations referred to in this report may be more clearly understood, I have thought it best to commence with the upper coal measures as exposed along the Missouri River near Omaha and the mouth of the Platte.

Omaha, which is well known to be the eastern terminus of the Union Pacific railroad, is built upon the northwestern rim of the coal measures as seen along the Lower Missouri. These rocks occupy a considerable portion of the State south of the Platte River, but north of that point they cover only a small portion of Sarpy and Douglas Counties. The last exposure of any importance is near the point decided upon as the location for the railroad bridge across the Missouri. The limestones at this point have been quarried for many years, but the amount of labor required to remove the vast thickness of marl and drift above it, will diminish greatly the importance of this quarry. Near Florence these limestones are seen in the bottom of the river at very low water, and near De Soto, obscure exposures have been detected. From that point to the foot of the mountains these rocks are not again seen. Along the Platte River for about eight miles there are extensive quarries of limestone that are very useful for building purposes. Scattered over the surface of the country in the two counties of Douglas and Sarpy, are exposures of the rusty sandstone of the Dakota group; and at the mouth of the Elkhorn River all traces of the coal measure rocks have disappeared, and do not reappear again until we reach the very margin of the mountains, over five hundred miles to the westward. After leaving the mouth of the Elkhorn very few exposures of rocks are seen for the next hundred miles, but there are enough to show that the underlying rocks are of cretaceous age. Near the mouth of Elkhorn River the sandstones of the Dakota group are seen, while on the distant hills traces of the yellow, chalky limestone, No. 3, occur. After reaching a point along the Platte about one hundred miles west of Omaha, the light, yellowish marls and sands of the White River group overlap the older rocks and occupy the country to the very margins of the Rocky Mountains. But the most important formation, and one that has a more favorable influence on the State of Nebraska than any other, is of very recent date in geological history. In the valley of the Missouri River, extending up nearly to Fort Pierre, and also to the mouth of the Missouri, and probably southward to the valley of Mexico, is a deposit of yellow marl varying in thickness from a few feet to several hundred. It has been called "the bluff formation," for it constitutes the picturesque bluffs or high hills which form the most conspicuous features in the scenery along the Missouri River. This yellow marl also enters largely into the composition of the soil of the vast bottom lands of the river which are so justly celebrated for their fertility. It is, however, in the immediate proximity to the water-courses that this yellow marl deposit is the thickest, and it gradually diminishes in depth as we recede from them; still, it is to this deposit that a very large portion of the West is indebted for its unsurpassed fertility and productiveness. It covers the country with such uniformity that it conceals almost entirely the basis rocks from view. Underlying this marl is a considerable deposit of drift material, as rounded pebbles or boulders and coarse sand, often presenting the most singular illustrations of oblique layers of deposit. The marl is usually quite homogeneous in its composition, and almost or entirely destitute of stratification, and the materials seemed to have been deposited in very quiet waters, and to have settled to the bottom of a fresh-water lake like gently-falling snow. The drift materials, as a rule, exhibit the irregular laminæ as if they had been deposited by currents of water. The exceedingly great importance of this yellow marl deposit is not yet well understood or appreciated, but it seems to me that the wonderful fertility of the soil of the western States and Territories, and its permanent productiveness for all time to come, is due to it.

The eastern portion of Nebraska is already quite thickly settled, and is susceptible of cultivation, but the western part must be inhabited, if settled at all, by a pastoral people.

These broad, level prairies are covered with a thick growth of short, nutritious grass, but the scarcity of water for the purpose of irrigation, and the almost entire absence of forest trees, must ever prevent settlements to any great extent. In the autumn nearly all the smaller streams dry up entirely, and several seasons the Platte has been known to become so low as to have no continuous current. It is a peculiar feature of these western streams, at times to be larger toward their sources than at their mouths. The Platte in its various branches always has an abundant supply of water, as their heads issue from the mountain sides, but in traversing the plains there are few or no springs or branches entering into it, or the water is entirely absorbed by the arid earth or thirsty air, until the bed becomes as dry as the dusty road. Hence all over the Rocky Mountain regions in the autumn are what are called dry creeks, with beds which, when full in the spring time, form large rivers.

The Platte River flows, for a distance of over four hundred miles, through the southern portion of what I have termed the White River tertiary basin, in contradistinction to the great lignite tertiary basin. The former has been separated into two formations, the White River group and the Loup River beds, on account of the organic remains characterizing each. The two former are entirely distinct, not a species passing from one to the other. I have supposed hitherto that the Platte River flowed through strata belonging to the Loup River group. They are certainly of quite recent age, but the pliocene remains that I collected on the Niobrara River came from loose gray sands which rested with a certain kind of unconformability on the eroded surface of the White River group. It is plain also that the valleys of the more important streams have been worn out, to some extent, prior to the deposition of the pliocene sands.

In the valley of the Niobrara and Loup Fork the pliocene sands are quite thick, and the line of separation between them and the White River group is very irregular, while on the hills the sands occur in many places, on and in, isolated hills.

The details of the geology of this most interesting region still remain to be worked out, and its geographical extent will be found to be much larger than has hitherto been supposed. The soil composed of the eroded materials of this basin is of moderate fertility, but owing to a want of water cannot be cultivated to any great extent. The greater portion of the surface underlaid by these beds is covered with a fine growth of grass which is especially adapted to the raising of sheep, and I am glad to see that some enterprising persons are making the experiment. The healthfulness of the climate, the nutritious character of the short grass, and the dryness of the ground, not unfrequently covered with small pebbles, must act favorably on sheep.

That portion of Wyoming east of the Laramie range, and south of the line of the Union Pacific railroad, is entirely covered with the upper beds of the White River tertiary basin. The valley of Lodge Pole, Crow Creek, and Chugwater, show the formations of this basin very distinctly from mouth to source. The Union Pacific railroad ascends the eastern slope of the Laramie range on a sort of bench of this formation, which seems to be unusually developed, and to extend without much interruption up to the very margin of the mountains, sometimes concealing all the rocks of intermediate age and resting on the syenites.

About twenty miles south of Cheyenne these beds disappear entirely along the eastern flanks of the mountains, and the lignite tertiary beds are exposed to view.

CHAPTER I.

FROM CHEYENNE TO DENVER.

I commenced my labors at Cheyenne, Wyoming Territory, and proceeded southward along the eastern flanks of the Rocky Mountains. My preliminary report will be little more than a transcript of my journal from day to day. It will be, therefore, impossible to systematize it as I would wish, or avoid in many cases repetition. There is great uniformity in the geology of the country, and when one has become familiar with the different geological formations over a small area, he can trace them with great rapidity over long distances. This will account, in part, for the large extent of country which I have been able to examine in a single season. The geological formations immediately underlying Cheyenne are of tertiary age, probably pliocene or very late miocene. The beds have been slightly disturbed by the upheaval of the mountain range, but their position in relation to the older tertiary beds shows their deposition to have been of late date. They are found deposited in the valleys and sometimes high on the mountain sides, and it is very seldom that they dip at an angle of more than five degrees. These beds can be traced far northward to the Black Hills of Dakota, a distance of three hundred and fifty miles, and they are thus shown probably to be the upper beds or most recent formation of the White River tertiary. Along the base of the mountains the rocks are mostly pudding-stone, or an aggregate of small water-worn pebbles, mostly very small, but sometimes several inches in diameter. These pebbles grow smaller and fewer in quantity as we recede from the mountains until they entirely disappear, and fine sand or marl takes their place. Near Cheyenne there is a bed of fresh-water limestone which is much used as lime, and seems to answer an excellent purpose in mason work and for whitewashing, and I have no doubt that such beds or layers occur in this basin everywhere. Along the line of the Union Pacific railroad, just before reaching Granite Cañon, a bed of the most excellent limestone crops out, on the margin of the range, of carboniferous age. This is burned into lime of snowy whiteness and is a great favorite with masons. It contains some fossils of well known carboniferous forms, as *Athyris subtilita*, *Productus pratteniana*, and crinoidal fragments. The red sandstones are exposed in a narrow belt along the margins of the mountains, but all rocks of more recent date are concealed by the tertiary deposits. In order that I may make my description of the different formations in their southern extension more clearly understood, I will describe them in as brief a manner as possible, as they have been studied in the regions to the northward.

The granites and metamorphic rocks do not differ in many respects from those which form the nucleus of the mountain ranges generally. Reddish and gray granites form the central portions, and on the sides is a series of stratified metamorphic rocks of a great variety of structure and composition. At the north the igneous rocks do not seem to predominate in the eastern ranges, but as we proceed southward toward New Mexico they increase in extent and force.

The Potsdam sandstone is the only member of the silurian that I have ever observed along the margins of the mountains. It was first discovered west of the Missouri River in the summer of 1857, during the exploration of the Black Hills of Dakota, by a United States expedition under the command of General G. K. Warren, United States Army, and it has been observed in several other localities since that time.

The following section of the Potsdam sandstone in its relation to the carboniferous beds, as observed by me around the margins of the Black Hills, shows the typical characters of each, where they are well exposed and have been clearly identified by organic remains:

1. Hard, compact, fine-grained, yellowish limestone of an excellent quality; passing down into a yellow calcareous sandstone, quite friable. Fossils: *Rhynconella rocky-montana*, *Athyris subtilita*, *Cyrtoceras*, *&c.*—50 feet.

2. Loose layers of very hard yellow arenaceous limestone with a reddish tinge, underlaid by a bed, six or eight feet in thickness, of a very hard blue limestone. The whole contains great quantities of broken crinoidal remains with cyathopylloid corals and several species of brachiopoda—40 feet.

3. Variegated sandstone of a gray and ferruginous reddish color, composed chiefly of grains of quartz and particles of mica, cemented with calcareous matter. Some portions of the bed are very hard, compact, siliceous; others a coarse friable grit; others conglomerate. Fossils: *Lingula prima*, *L. antiqua*, *Obolella nana*, and *Arionellus oweni*—50 feet.

4. Stratified metamorphic rocks in a vertical position for the most part.

Rocks about the same as those above described, sometimes very much thicker and sometimes thinner, have been seen, more or less, all along the margin of the Rocky Mountains, on both sides the main axis from the north line to Cheyenne.

About the sources of the Missouri River, along the flanks of the Big Horn and Wind River Mountains, these rocks are particularly developed. Now and then they all disappear for a considerable distance, and then, at the first favorable opportunity, reappear from beneath beds of more recent date. A series of arenaceous beds, which we have called the "red arenaceous deposits, or triassic," form one of the most conspicuous features of the geology along the flanks of both sides of the principle ranges of mountains and are almost always present. They were first observed by me, forming a narrow belt or girdle around the granite nucleus of the Black Hills of Dakota, in the summer of 1857. These rocks are sometimes called saliferous or gypsum-bearing beds, from the fact that they contain both salt and gypsum, the latter mineral oftentimes in great quantities. There are also mingled with these beds several layers of bluish siliceous limestone, which at the far north attain a considerable thickness, but southward thin out to a few feet, or are entirely concealed by the debris which everywhere prevails.

These red beds, when they make their appearance, often give the most unique and remarkable features to the scenery, and any development of them, however small, never fails to attract even the commonest observer on account of their brick-red appearance. No well-authenticated fossils have ever been found in them, yet they are regarded as of triassic age by the common consent of geologists. I am inclined to believe that a portion of the upper light-red beds, with the included layers of flinty limestone, are jurassic, but I have never been able to find any well defined line of separation between what are well known to be jurassic and the supposed triassic beds.

Resting above these red beds is a series of marls and arenaceous marls of a light or ashen gray color, with harder layers of limestone or fine sandstone, which were also first discovered around the margin of the Black Hills of Dakota in 1857. Since the discovery in the Black Hills, jurassic fossils have been found over a very wide geographical area, and yet I have never seen them so well developed, or the peculiar fossils so abundant, as at the locality where they were first observed. Although I have traced this jurassic belt by its organic remains over many hundreds of miles, I have been able to discover scarcely a well-defined jurassic fossil south of Deer Creek, a point one hundred miles north of Fort Laramie, or south of the Lake Como, on the Union Pacific railroad.

I believe that a thin remnant of this belt extends far south to New Mexico, but it is often so obscured, or so easily concealed, that I have been continually in doubt in regard to its existence. Coextensive with all the mountain ranges is a large series of beds above the jurassic belt which belong to the cretaceous period, the upper and middle portions of which are everywhere indicated by characteristic fossil remains, as seen on the Missouri River, where they were first studied by Mr. F. B. Meek and the writer. The cretaceous rocks present five well-marked divisions, Nos. 1, 2, 3, 4, and 5, or Dakota group, Fort Benton group, Niobrara division, Fort Pierre group, and Fox Hill beds. On the Lower Missouri No. 1, or Dakota group, is characterized by several species of marine shells and a profusion of impressions of deciduous leaves; but along the margins of the mountain elevations I have never been able to discover a single specimen of organic remains that would establish the age of the rocks. I only know that there is a series of beds of remarkable persistency all along the margin of the mountain ranges, holding a position between well-defined cretaceous No. 2 and jurassic beds, and in my previous reports I have called them transition beds, or No. 1. They consist of a series of layers of yellow and gray, more or less fine-grained sandstones and pudding-stones, with some intercalated layers of arenaceous clays. In almost all cases there is associated with these beds a thin series of carbonaceous clays, which sometimes becomes impure coal, and contains masses of silicified wood, &c. On the west side of the Black Hills they assume a singularly massive appearance, nearly horizontal, two hundred to two hundred and fifty feet thick, and are called Fortification Rocks. Here also occurs a thin bed of carbonaceous clay. On the eastern slope of the Big Horn Mountains I observed this same series of beds in the summer of 1859, holding a position between cretaceous No. 2 and the jurassic marls, with a considerable thickness of earthy lignite, large quantities of petrified wood, and numerous large uncharacteristic bones, which Dr. Leidy regarded as belonging to some huge saurian.

There are very few points of resemblance between these beds and those which form the Dakota group, as seen in Kansas and Nebraska.

All the evidence therefore that I have had to guide me in regard to these beds along the margin of the mountain ranges has been their position.

No. 2, on the Missouri River, is composed of very black plastic clays, with some thin layers of limestone and sandstone, and is quite well separated from No. 1 below and No. 3 above. No. 3 is composed of massive layers of chalky limestone, always containing *Inoceramus problematicus* and *Ostrea congesta.*

Along the Kansas Pacific railroad, at Hayes City and Fort Wallace, No. 3 occurs in such massive layers that it is sawed into building blocks with a common saw. No. 4 is a dark ashen steel-colored laminated clay, with bluish calcareous concretions filled with shells. No. 5 is a yellowish ferruginous arenaceous clay, with the greatest abundance of molluscous fossils. At various localities all along the margin of the mountain ranges these divisions of the cretaceous are far less distinctly separated, and vary more or less in their structure and composition, and yet in tracing them carefully and continuously from the Missouri River they always retain enough of their typical character, so that I have never been at a loss to detect their presence at once, although after leaving the Missouri River we do not find any well-defined lines of separation, either lithologically or paleontologically.

With the commencement of the tertiary was ushered in the dawn of the great lake period of the West. The evidence seems to point to the conclusion that from the dawn of the tertiary period, even up to the commencement of the present, there was a continuous series of fresh-water lakes all over the continent west of the Mississippi River. Assuming the position that all the physical changes were slow, progressive, and long-continued, and that the earlier sediments of the tertiary were marine, then brackish, then purely fresh water, we have through them a portion of the consecutive history of the growth of the western continent, step by step, up to the present time. The earliest of these great lakes marked the commencement of the tertiary period, and seems to have covered a very large portion of the American continent west of the Mississippi, from the Arctic Sea to the Isthmus of Darien.

As I have before stated, the first sediments were marine, then came brackish water, and soon purely fresh water, as is plainly indicated by the organic remains. The lower beds of the great lignite basin everywhere contain layers, varying from a few inches to two feet in thickness, made up almost entirely of oyster shells, with a few other species of marine or estuary types. No exclusively marine forms have as yet been found to my knowledge, but as we ascend in the beds all traces of the salt sea disappear, and a great profusion of fresh-water and land shells appear, with vast quantities of the impressions of leaves of deciduous trees. Numerous beds of coal, varying in thickness from a few inches to fifteen or twenty feet, characterize this deposit.

About the middle of the tertiary period the second extensive lake commenced in the West, which we have called the White River tertiary basin. We believe that it commenced its growth near the southeastern base of the Black Hills, and gradually enlarged its borders. I am inclined to think that this lake has continued on, almost or quite up to the commencement of the present period; that the light colored arenaceous and marly deposits in the Park of the Upper Arkansas, in the Middle Park, among the mountains at the source of the Missouri River, in Texas and California, and Utah, are all later portions of this great lake. The upper miocene or pliocene deposits in the Wind River Valley, near Fort Bridger, and on the divide between the Platte and the Arkansas Rivers, were undoubtedly synchronous, though perhaps not connected with this great basin. Every year, as the limits of my explorations are extended in any direction, I find evidences of what appear to be separate lake basins, covering greater or less areas, and bearing intrinsic proof, more or less conclusive, of the time of their existence. I have given in this place the above brief description of the various geological formations as I have studied them in the West, in order that my subsequent remarks on these formations in their southern extension may be more clearly understood. Constant reference will be made to rocks as they have been seen in the far North and West, in order that the story of their geological extension may be linked together.

June 29, 1869.—Left Fort D. A. Russell about 10 o'clock in the morning with my entire party, consisting of twelve persons and eighteen mules and horses, with two large covered wagons and an ambulance. By the kindness of Colonel E. B. Carling, the depot quartermaster, at Fort D. A. Russell, I was provided with everything needful for independent camp life, and I at once commenced my explorations in earnest.

We traveled to-day thirteen miles southward from Fort Russell. Our entire route was over the more recent beds of the White River tertiary basin. The lowest bed exposed by the cuts in the streams, is a thick layer of flesh-colored indurated marl, much like that containing so many vertebrate fossils on White River, Dakota. It contains some thin layers of very fine gritty rock. Overlying this is a thick bed which appears more recent, yet apparently conforms to the marl beds below. It is composed of water-worn pebbles of various sizes, forming a real pudding-stone. Near the margins of the mountains this bed gives the characteristic features to the scenery, as it is cut through by the myriad small streams that issue from the mountain side. It is at least three or four feet in thickness. Most of the pebbles are from the granite rocks that form the central portions of the Laramie range. The beds all dip from the mountains eastward at a moderate angle, and it is evident that this entire formation was deposited after the mountain ranges had nearly reached their present height. The strata seldom dip at an angle of over 5° and rest unconformably on the older beds when they are seen in apposition. Near the junction of the metamorphic rocks with these modern pudding-stones the pebbles or bowlders are not much worn, and of moderate size, six to twelve inches in diameter, but the sediments grow finer and finer as we recede from the foot of the mountains until the pudding-stones pass into a fine-grained whitish sandstone. We can see, therefore, that these deposits formed the proper rim of the fresh-water lake, that the sediments were derived from the erosion of the feldspathic granites, and that the forces that were in operation acted from the direction of the mountain ranges.

There are also vast quantities of drift material which I regard as local. It seems to me that the evidence is clear that all this modern drift-action had its origin in the mountain ranges in the immediate vicinity; that in earlier times the snow and ice gathered on the summits in vastly greater quantities, and that in melting, from year to year, in the form of water and ice, they brought along vast quantities of rocks from the mountains and distributed them over the surface.

The waters, with the masses of ice, would naturally follow the channels of the streams if they had been marked out, or they would mark out new channels, for nearly or quite all the valleys that extend down from the mountains become shallower, the further they extend eastward from the flanks of the range. This superficial deposit at the very margins of the mountains is composed of very coarse materials, sometimes immense masses of granite of all kinds, but slightly worn; but proceeding from the base of the mountains, the rocks become smaller and more rounded, until they pass into small pebbles, mingled with loose sand.

The phenomena of erosion, as seen at the present time, all along the flanks of the mountains, in the plains, in the channels of streams, point clearly to a vastly greater quantity and force of water than exist anywhere at the present time.

The surface of the country along the base of the mountains is extremely undulating—worn into hill, valley, ridge, or rounded buttes. The strata in these ridges and hills show that the entire surface was much higher than it is at present, and that these ridges and buttes are only remnants of beds left after the erosion, and how great a thickness of strata was originally deposited above these remnants, and is now entirely swept away, it is impossible to determine, though we believe it was very great.

Now, on these hills are the greatest numbers of large, rounded stones, of all kinds, granite and sedimentary, as if they had been left there by the melting masses of ice which had lodged on the hills. These stones are also accumulated in long lines or belts, as if they had been driven by currents so as to form shore lines, or lodged-in eddies. The evidence is clear that great bodies of water, in which were probably mingled masses of ice, swept over the plain country within a comparatively recent geological period.

Opposite Camp Carling, in the bluffs of Crow Creek, a good thickness of drift is seen filling up the irregular surface of the modern tertiary beds, so that we have evidence of quite extensive erosion of the surface prior to the deposition of this drift.

Along all the main water-courses are high ridges showing the rocky strata perfectly. A little lower is a second ridge, mostly grassed over, but more or less parallel with the higher ridge; then we have a graduated series of terraces, from one to three, extending down to the water's edge. This description applies to all the main water-courses along the base of the mountains, whether there is running water in them at this time or not; and they all seem to give evidence that they once contained far more water than at present. This configuration of the surface aids much in giving a sort of picturesque appearance to the plains, inasmuch as we cross these undulations at right angles in traveling north to south.

The soil in the valleys of the streams is rich enough, and when it can be irrigated, will produce good crops; and not until the farmers and stock-growers begin to settle about Cheyenne will it have a permanent and substantial growth.

June 30. The distance from Cheyenne to Laporte, on the Cache la Poudre, is forty miles. The tertiary pudding-stone beds extend along the immediate flanks of the mountains for twenty-five miles, but disappear from the plains within ten or fifteen miles of Laporte.

I have estimated their entire thickness here at from twelve hundred to fifteen hundred feet. The high hills near the station are capped with coarse sandstone, with horizontal strata, and are eight hundred feet above the bed of the creek that flows near their base. From beneath these recent beds arise the more sombre-hued beds of the lignite tertiary. We have then broad grassy plains, dotted here and there with buttes like truncated cones, and long narrow belts of table-lands, with perfectly plain surfaces to the eye, from a distance. Why these more modern tertiary beds are so persistent along the immediate sides of the mountain, but have been entirely swept away ten miles to the eastward, I cannot tell. This narrow belt, about ten or fifteen miles wide, extending up to the granite rocks, and for the most part concealing all the intermediate rocks, forms a sort of bench, with a gently ascending grade for the Union Pacific railroad.

Either above or below this bench the ascent to the mountains would have been very difficult, expensive, and perhaps impossible.

About twenty miles south of Cheyenne a bed of coal has been opened and wrought to some extent. The outcropping revealed a bed of impure coal four feet eight inches thick, with an inclination 12° east. The coal became more valuable as it was worked further into the earth; and by following the direction of the dip, the coal was found to be five feet four inches thick. In nearly all instances coal beds increase in thickness, rather than diminish, the further they are explored. The whole plain country is covered with such a thickness of superficial drift that it is next to impossible to obtain a connected section of the underlying rocks. Sometimes a stream will cut so deep that a portion of them is exposed, and by following it a great distance, the order of superposition may be obtained with some degree of correctness.

A section across the upturned edges of the strata, from the direction of the mountains eastward, is as follows:

7. Drab clay passing up into areno-calcareous grits composed of an aggregation of oyster-shells, *Ostrea subtrigonalis*.
6. Lignite—5 to 6 feet.
5. Drab clay—4 to 6 feet.
4. Reddish, rusty sandstone, in thin laminæ—20 feet.
3. Drab arenaceous clay, indurated—25 feet.
2. Massive sandstone—50 feet.
1. No. 5, cretaceous apparently passing up into a yellowish sandstone.

The summits of the hills near this bed of coal are covered with loose oyster shells, and a stratum four feet thick, or more, is composed of an aggregation of them. This species seems to be identical with the one found in a similar geological position in the lower lignite beds of the Upper Missouri, near Fort Clark; also at the mouth of the Judith, and at South Boulder, and doubtless was an inhabitant of the brackish waters which must have existed about the dawn of the tertiary period in the west. It would seem, that in these lower coal beds the molluscous life was almost entirely confined to this genus, (from three to five species having already been discovered.) Near Medicine Bow Creek there is a thin seam of oyster shells, quite minute, and at Point of Rocks, on the Union Pacific railroad, above several beds of coal, there is a layer two feet or more in thickness, made up of the shells of a fine large species, about the size of our common edible oyster.

On the Upper Missouri a great variety of well-known fresh-water types of shells are found in the strata connected with the coal, especially toward the middle portion. But southward I have never met with any other shells than oysters, in direct connection with the coal beds.

During the summer of 1859 I traced this lignite formation uninterruptedly from the Upper Missouri Valley to a point on the North Platte, about eighty miles northwest of Fort Laramie. It is there overlapped by the modern tertiary deposits previously described, but reappears about twenty miles south of Cheyenne, and extends with some interruptions far southward into New Mexico. During the summer of 1868 I traced these coal beds, on the other side of the mountains, westward nearly to Salt Lake City; and in the Middle and South Parks there are quite extensive developments of them.

I make these remarks as confirmatory of a statement which I made in an article in Silliman's Journal, March, 1868, "that all the lignite tertiary beds of the west are but fragments of one great basin interrupted here and there by the upheaval of mountain chains, or concealed by the deposition of newer formations."

As soon as the lignite beds reappear southward the aspect of the country changes. The distant hills that flank the mountains on the right are still the pebbly conglomerate beds. And in the valleys of the little streams, about four miles south of Spotswood Springs, are several exposures of beds which are undoubtedly older tertiary. There is here shown a deep yellow arenaceous indurated clay layer, passing down into an ashen-brown grit, with rusty yellow concretions. All over the hills are scattered the greatest number of water-worn boulders. The lignite strata incline in the same direction as those of the more modern deposits. The dip of the former is about five degrees to ten degrees, the latter one degree to three degrees east, from the mountains. There are many other localities where the evidence of non-conformity of the two deposits is perfectly clear.

A little further eastward on the Dry Creek the ridge is capped with gray, loosely laminated sandstone; while in the indurated arenaceous bed below are beds of massive rusty sandstone, the same as those that compose the natural fortifications about thirteen miles southwest of Cheyenne. The ridge extends far eastward into the plain, with the beds nearly horizontal.

Near a high conical butte a little further southward we find the lignite beds dipping 85° with a strike nearly north and south. And in the south and southwest we can see the upturned ridges of cretaceous and older sedimentary formations composing the flanks of the mountains. The modern tertiary and the superficial drift deposits have been so removed from the mountain side—about ten or fifteen miles north of Cache la Poudre—that all the unchanged sedimentary rocks in this region are revealed in the form of inclined ridges, which gradually die out in the plains eastward like sea waves.

A bed of the laminated chalky marl of No. 3 with *Ostrea congesta* and *Inoceramus problematicus* is particularly noticeable. In the lignite beds harder layers of rusty sandstone, with loosely laminated arenaceous clay, and the softer materials are worn away by erosion, leaving the harder rocky layers projecting above the surface in long lines like walls.

Near Park station, about twelve miles north of Cache la Poudre, the upheaved ridges begin to spread out, revealing very clearly to the scrutiny of the geologist all the sedimentary rocks, to the tertiary inclusive. Commencing in the plains about ten miles east of the margin of the mountains we find a series of gently inclined tertiary sandstones, dipping from 5° to 10°. Then come the complete series of cretaceous strata in their order, inclining from 20° to 35°. Underneath the ridge capped with the sandstone No. 1 is a thin belt of ashen-gray marls and arenaceous marls, with one or two layers, two to four feet thick, of hard blue limestone, which I regard as of jurassic age. These pass down into light reddish, loose arenaceous sediments. Further toward the mountains, come one to three ridges of brick-red sandstone, and loose red sandy layers, sometimes variegated. Close to the margin of the mountains, sometimes forming the inside ridge, is a bed of whitish limestone, underlaid by dull purplish sandstone and pudding-stones, which are probably of carboniferous age. These beds dip at various angles, from 30° to 60°, and, as far as I can determine, conform generally to the inclination of the metamorphic rocks which compose the mountain nucleus.

The opening in the foot-hills of the mountains through which Box Elder Creek flows exhibits the red beds and jurassic in full development. The whitish-gray sandstones, which lie between the red beds and the well-marked cretaceous strata, contribute much toward giving sharpness of outline to the hills, and the broken masses of rock from this bed are scattered over their sides.

The valley of the Box Elder is very beautiful, and, like the valleys of most of the little streams here, makes its way through the ridges and flanks of the mountains, nearly at right angles to the trend of the strata.

All these ridges, or "hog-backs," as they are called by the settlers of the country, vary much in the angle of dip. It not unfrequently occurs that the outer and more recent ridges incline at a very high angle, or stand nearly vertical; and there are many examples where they have been tipped several degrees past verticality; while the inner sandstone ridges, lying almost against the metamorphic rocks, incline at a small angle or are nearly horizontal; and again this may be reversed. These mountain valleys are not only beautiful, but they are covered with excellent grass, making the finest pasture grounds for stock in the world. The animals are so sheltered by the lofty rock-walls on each side that they remain all winter in good condition without any further provision for them.

The Box Elder separates into two branches in the foot-hills, and between the forks there is a large circular cone with nearly horizontal strata of the red beds. A section, ascending, would be as follows:

1. Brick-red sandstone with irregular laminæ and all the usual signs of currents or shallow water. Some of the layers are more loosely laminated than others, thus causing projecting portions—300 to 400 feet.

2. The red sandstone passes up into a yellow or reddish-yellow sandstone, massive—60 feet.

3. Passing up into a bed of grayish yellow rather massive sandstone—50 feet.

4. Ashen-brown nodular or indurated clay, with deep, dull purple bands; with some layers of brown and yellow fine-grained sandstone, undoubtedly the usual jurassic beds with all the lithological characters as seen near Lake Como, on the Union Pacific railroad—150 to 200 feet.

Near the base of these beds are thin layers of a fine grained grayish calcareous sandstone, with a species of *Ostrea* and fragments of *Pentacrinus asteriscus*. Scattered through this bed are layers or nodules of impure limestone.

5. Above this marly clay there is at least two hundred feet of sandstone and laminated arenaceous material, varying in color from a dirty brown to grayish white, with layers of fine grayish-white standstone.

I do not hesitate to regard the beds described as 4 and 5 as of jurassic age, and they are better shown here than at any other point between Fort Laramie and the south line of Colorado on the eastern slope of the Rocky Mountains. Usually the most abundant and most characteristic fossil in the jurassic beds, when exposed, is *Belemnites densus*, but that has not been observed south of Lake Como, west of the Laramie range. As we proceed southward these jurassic beds become thinner and more obscure, so that it often becomes a matter of doubt whether they exist at all.

We have, also, in this vicinity an illustration of the difference of inclination in the same series of upheaved ridges. In the plains some of the lower lignite tertiary beds and cretaceous No. 5 stand nearly vertical, or 85° east. No. 4 fills the intervening valley with its dark shale, and the next ridge west—cretaceous No. 3—inclines 36°. Then come the jurassic beds capped with the sandstones of No. 1, inclining 8°. Then comes a series of red beds dipping 1° to 3°. The inner ridge, or "hog-back," is the largest of all—one hundred and fifty to two hundred feet high—is partly covered on the east, or sloping side, with the loose red sand of the triassic; and on the west or abrupt side, is revealed a considerable thickness of limestone, which I suppose to be of carboniferous age. This ridge is remarkably furrowed on the eastern slope by streams, but is too high up on the mountain side to be divided by the currents into the peculiar conical fragments, as the lower ridges are. And hence it presents an almost unbroken flank for miles. There is no better exhibition of the sedimentary rocks, with all their peculiar characteristics and irregularities, than from the head of Box Elder Creek to Cache à la Poudre, where the belt of upheaved sedimentary rocks varies from five to fifteen miles in width. No one could stand on the summit of one of these ridges and turn his eye westward over the series, rising like steps to the mountain summit, and then looking eastward across the broad level plain where the smaller ridges die out in the prairies, like waves of the sea, without arriving at once to a clear conception of the plan of the elevation of the Rocky Mountain range.

The main range of the mountains is really a gigantic anticlinal, and all the lower ranges and ridges are only monoclinals, descending, step-like, to the plains on each side of the central axis. There are some variations from this rule at many localities, which I shall attempt to explain from time to time in the proper place.

One of these ridges is quite conspicuous to the eye, from the fact that it is capped with a heavy bed of sandstone which I have always regarded as transition or No. 1 (?) because it holds a position between the well-defined cretaceous beds Nos. 2 and 3, and the jurassic.

Not a single well-marked fossil, animal or vegetable, has ever been found in this group of strata along the flanks of the mountains; yet I do not hesitate to regard them as lower cretaceous.

On the summits of all these ridges are numerous piles of rocks which have been erected by Indians in years past as monuments or land-marks.

Inside of the sedimentary ridges are the metamorphic rocks, mostly red feldspathic granites, disintegrating readily, and easily detected by the eye at a distance by their style of weathering. Still further westward are the lofty snow-capped ranges, whose eternal snows form the sources of the permanent streams of the country.

It seems clear to me that the more recent sedimentary formations, up to the lignite tertiary, inclusive, once extended over the whole country. Perhaps no finer locality exists in the West for the careful study of the different sedimentary formations and their relations to the metamorphic rocks than along the overland stage road from Laramie to Denver.

Before reaching Laporte the road passes for twenty miles or more through ridge after ridge remarkably well exposed. After emerging from the mountains eastward it runs south for four or five miles along the cretaceous beds with their upturned edges on the east side, and the jurassic and triassic (?) on the west forming a slope much like the roof of a house. The valley between the two ridges through which the road runs is a beautiful one.

South of Big Thompson Creek the belt of upheaved ridges, or unchanged rocks, becomes quite narrow, and continues so to Denver, and even beyond.

The cretaceous rocks in this region, though plain to one who has carefully studied them on the Upper Missouri, are not separated into well-marked divisions. If they had first been studied along the foot of the mountains only from Cheyenne southward, it is very doubtful whether the five distinct groups of strata would have been made out. The three divisions, upper, middle, and lower cretaceous, are more natural south of the North Platte, inasmuch as Nos. 2, 3, 4, and 5 pass into each other by imperceptible gradations. Though very few organic remains are observed in them, yet I have never found the slightest difficulty in detecting the different divisions at a glance by their lithological characters, but I find it quite impossible to draw any line of separation that will be permanent. Quite marked changes occur in the sediments of these divisions in different parts of the West, but by following them continuously, in every direction, from their typical appearance on the Upper Missouri, the changes are so gradual that I have never lost sight of them for a mile, unless concealed by more recent deposits.

As I have before stated, I regard the group of sandstones which are always found between well-defined cretaceous No. 2 and the jurassic beds as No. 1, or transition. No. 2 is certainly well shown, with many of its features, but it is a black shale, often arenaceous, containing many layers of sandstone with some concretions; but so gradually passing up into No. 3 that it is quite impossible to separate the two. Only in thin portions of either Nos. 2 or 3 do we find any resemblance to the same groups as shown on the Upper Missouri. No. 3 is a thinly-laminated yellow chalky shale with some layers of gray, rather chalky limestone, always containing an abundance of *Inoceramus*, doubtless *I. problematicus*, and *Ostrea congesta*. Remains of fishes are almost always found in the dark shales of No. 2. The black shales of No. 4 are quite conspicuous and well marked, and have been quite thoroughly prospected for coal, but to no purpose. These black shales pass gradually up into yellow rusty arenaceous clays which characterize No. 5; and No. 5 passes up into the lignite tertiary beds, where they can be seen in contact, without any well-defined line of separation that I could ever discover. But few species of fossils are found in Nos. 4 and 5 in their southern extension, but *Baculites ovatus* and several species of *Inoceramus Ammonites*, &c., are common. Another feature is well marked here, and that is, there are no beds that indicate long periods of quiet deposition of the sediments. Nearly all the sediments indicate either comparatively shallow water or currents more or less rapid.

Sometimes a single ridge will include all the beds of one formation, or even those of two or three. I have often seen the sandstones of No. 1, the jurassic, and a portion of the triassic included in one ridge and the adjoining valley. Again, a single formation will be split up into two or more ridges.

On the Cache à la Poudre, about a mile above Laporte, on the south side of the river, the sandstones of No. 1 are separated into four successive ridges, inclining, respectively, 18°, 21°, 35°, and 46° about southeast. Much of this sandstone is a fine-grained grayish white, and rusty yellow color, sometimes concretionary, or like indurated mud. Here all the divisions of the cretaceous extend eastward in low ridges until they die out in the plains or are concealed by the overlying tertiary. Along the Cache à la Poudre and its branches is a series of terraces which are quite uniform.

This valley is one of the most fertile in Colorado. The present year there has been so much rain that irrigation has been unnecessary. The bottom lands are about two miles wide, and thickly settled from mouth to source. The grass is unusually fine this year everywhere.

July 2.—In company with Dr. Smith, of Laporte, I visited the supposed gold and copper mines at or near the sources of the Cache à la Poudre River. This stream makes its way through what might be called a monoclinal rift, or between two ridges, whether of changed or unchanged rocks which incline in the same direction. We ascend to the axis of the main Rocky Mountain range by a series of step-like ridges, each one inclining in the same general direction at some angle, with their counterparts on the opposite side of the main axis. Speaking of these ridges locally, I have called them in this report monoclinal, from the fact, that as a rule their counterparts, although they have once existed on the west side of the range, are in most cases swept away. We passed up a beautiful valley with the red beds on our left, and a few remnants of the red beds and the metamorphic rocks on our right, for about fifteen miles. We then came to the red feldspathic granites, in which the mineral lodes are located. We first examined a local vein of black rock, in which hornblende predominates. It contains some mica and iron, so that it might be called a local outcrop of black hornblende syenite. Masses of it have a rusty look from the decomposition of the iron in the rock, and sometimes it is covered with an incrustation of common salt or potash. Iron in some form is one of the prominent constituents of all the rocks of this region, changed or unchanged. So far as I could determine, the inclination of the metamorphic rocks is in the same direction as the sedimentary. I have assumed the position that all the rocks of the West are, or were, stratified, and that where no lines of stratification can be seen, as in some of the massive granites, they have been obliterated by heat during their metamorphism. Therefore all the metamorphic rocks, whether stratified or massive, that form the nucleus of the Rocky Mountains, must have some angle of dip, equally with the sedimentary rocks. In many cases I have to be guided by the intercalated beds of mica or talcose slates. I am of the opinion that there are anticlinals and synclinals among the metamorphic rocks of this region, and that the mountain valleys are thus formed for the most part.

We examined a number of lodes which were moderately rich in copper. All the lodes have a trend about northeast and southwest, and are two to four feet wide, with well-defined walls. Much of the gangue rock is spongy like slag, owing to the decomposition of iron pyrites; and there are large masses of the casts of cubes, evidently cubes of iron pyrites. Our examinations were not very thorough, but I was not favorably impressed with this district as a rich mineral region. Some of the copper mines, at some future day, may yield a fair return, but it will be many years before the country will be built up by its mineral wealth.

July 3.—Our route to-day was along the flanks of the mountains, from Cache à la Poudre to Big Thompson Creek. Lying over the red beds and appearing to form a dividing line between the red beds and the ashen-gray marly clays above, is a well-defined bed of bluish semi-crystalline limestone, two to four feet thick, somewhat cherty, though susceptible of a high polish, too brittle and liable to fracture in any direction to be valuable for ornamental purposes—probably useful for lime only. I regard this as of jurassic age, although I was unable to find in it any well-marked organic remains. The same bed occurs in the Laramie plains, where it contains many fragments of crinoidal stems, which Professor Agassiz referred to the well-known jurassic genus *Apiocrinites*.

On the summit of the first main "hog-back" is a bed of massive sandstone, immense blocks of which have fallen down on the inner side of the ridge, adding much to the wildness as well as ruggedness of the scenery. These rocks are made up almost entirely of an aggregation of small water-worn pebbles. The layers of deposition are very irregular, inclining at various angles. This irregularity in the laminæ is a marked feature of this sandstone. It forms a portion of the group which I have called transition, or No. 1. They are certainly beds of passage from well-marked cretaceous to the jurassic, and the lower portion being almost invariably a pudding-stone, they may well mark the boundary between the two great periods. In many places along our route this group forms lofty perpendicular escarpments, varying from thirty to sixty feet in height, indicating a considerable thickness of the massive sandstone. For fifteen miles we can pass along behind this hog-back ridge parallel with the mountains, through a most beautiful valley with fine grass, and over an excellent natural road. On our left are the upturned edges of a ridge capped with No. 1, passing down into the limestone and ashen marly clays of the jurassic, with a few feet of the red sandstone at the base, while the valley, which may be three hundred to five hundred yards wide, is composed of the worn edges of the loose red beds of the triassic, and on our right are the variegated sands and sandstones of the formation.

South of Cache à la Poudre there seem to be but two principal ridges between the transition group No. 1 and the metamorphic rocks, although at times each one of these ridges will split up into a number of subordinate ridges which soon merge into the main ridge again. In most cases the inner ridge includes all the red beds proper, and there is a well-defined valley between it and the metamorphic rocks, but sometimes the sedimentary beds flank the immediate sides of the metamorphic ridge. Through these ridges are openings made by the little streams which issue from the mountain's side. Sometimes these openings are cut deep through to the water level, and at other times for only a few feet from the summit. Sometimes there is a stream of water flowing through them, but most of them are dry during the summer. These notches in the ridges occur every few hundred yards all along the foot of the mountains.

The cretaceous and tertiary beds generally form several low ridges which are not conspicuous. The principal ridge outside, next to the plains, is composed of the limestones of No. 3, which is smoothly rounded and covered with fragments or chips of limestone. Between this and the next ridge west, there is a beautiful concave valley about one-fourth of a mile wide. The line between the upper part and the foot of the ridge proper is most perfectly marked out by the grass. The east slope of this ridge is like the roof of a house, so steep that but little soil can attach to it, and in consequence of this it can sustain only thin grass and stinted shrubs. These ridges are sharp or rounded, depending upon the character of the rocks of which they are composed. Cretaceous formation, No. 3, yields so readily to atmospheric agencies, that the ridges composed of it are usually low and rounded, and paved with chipped fragments of the shell limestone. The harder sandstones give a sharpness of outline to the ridges which has earned for them the appellation of "hog-backs," by the inhabitants of the country. In No. 3 I found *Ostrea congesta* very abundant, and a species of *Inoceramus* identical with the one occurring in the limestone at South Boulder, and the same as the one figured by Hall in Frémont's Report, Plate IV, Fig. 2, and compared with *Inoceramus involutus*, (Sowerby,) page 310. The lower part of No. 3, containing the *Inoceramus*, is a gray marly limestone, which passes up into a yellow chalky shale, which weathers into a rusty yellow marl that gives wonderful fertility to the soil, while the dark shales of Nos. 2 and 4, as well as the rusty arenaceous clays of No. 5, are distinctly revealed at different localities. The light-colored chalky limestones of No. 3 are more conspicuous at all times along the foot-hills of the mountains, even to New Mexico, than any other portion of the cretaceous group.

The valley of Thompson Creek is very fertile, varying from half a mile to a mile in width, is filled up with settlers, and most of the land is under a high state of cultivation. The creek itself is one of the pure swift-flowing mountain streams which have their source in the very divide or summit of the water-shed, and are rendered permanent by the melting of the snows. All these mountain streams would furnish abundant water-power, most of them having a fall of thirty feet to the mile.

There seems to be a decided improvement in the soil as we go southward. The geological formations are the same, but the climate is more favorable.

On a terrace on the north side of Big Thompson Creek there is a bed of recent conglomerate, quite perfect, and belonging to the modern drift period. It is very coarse, and the worn boulders are held together by sesquioxide of iron. I note it here as an example of very recent conglomerate. There is much fine sand, and the rounded stones are exactly like those which pave the bottoms of streams. The thickness of this boulder deposit is considerable, and it seems to underlie the whole valley portion of the country.

The cretaceous beds of No. 3 pass down into a yellowish sandstone which forms a low ridge on the north side of Big Thompson Creek. Two or three low ridges of cretaceous appear east of this point, but die out in the prairie. This ridge inclines 15°, then comes a valley about one-fourth of a mile wide, and a second ridge of rusty reddish fine-grained sandstone, evidently No. 1, or the transition group. This ridge inclines 25°. Underlying the sandstone, which forms a large part of this ridge, we find the ashen-gray marly and arenaceous clays of the jurassic, including some thin beds of sandstone and one layer of limestone four to six feet thick, which has been much used for lime among the farmers. These beds pass down without any perceptible break into the light brick-red sandstones which form the next two ridges westward. These beds have a dip of 30°. About the middle of the red beds there is a layer of impure limestone standing nearly vertical 65°, two to four feet thick, which has also been used somewhat for lime. The next ridge west has a rather thick bed—ten to fifteen feet—of very rough impure limestone looking somewhat like very hard calcareous tufa. The intermediate beds are loose brick-red sands.

There is here a somewhat singular dynamic feature—a local anticlinal. One of the ridges flexes around from an east dip to a west dip, from the fact that one of the eastern ranges of mountains runs out in the prairie near this point, forming at the south end originally a sort of semi-quaquaversal, the erosive action having worn away the central portions. This forms a short anticlinal of about a mile in length. On the east side of the anticlinal valley the principal ridges are shown, including nearly all the red beds; and on the west side, only the upper portions of the red sandstones with the jurassic beds and the transition sandstones. The latter rocks form the nearly vertical wall in which is located a somewhat noted aperture called the "Bear's Church." In the west part of this anticlinal, within twenty feet of the brick-red sandstones, is a blue, brittle limestone layer about six feet thick, inclining seventy-eight degrees. This west portion of the anticlinal might be described across the upturned edges thus, commencing at the bottom:

1. Rather light brick-red sandstones in three layers—estimated 200 feet.

2. The red bed passes up into a massive reddish-gray rather fine sandstone—20 feet.

3. Then comes a thin layer of fine bluish-brown sandstone—2 feet; then the bluish limestone—4 feet.

4. Then about twenty-five feet of ashen clay, with six to ten feet of blue cherty limestone, with some partings of clay.

5. About two hundred feet of variegated clays.

6. A bed of quite pure limestone, blue, semi-crystalline—four to eight feet. The grass prevents definite measurements, and all the beds vary in thickness in different places, as well as in dip, which is from 60° to 80°.

7. This intermediate space is covered over with a loose drab yellow sand, doubtless derived from the erosion of the edges of the beds beneath, which are supposed to be jurassic. There is one bed of limestone about two feet thick, similar to that before described. All these limestones appear to contain obscure fragments of organic remains.

8. A nearly vertical wall of sandstone; dip 60° to 65°. This bed is formed of massive layers, in all, one hundred and fifty feet thick or more, and is composed largely of an aggregate of small water-worn pebbles of all kinds. Most of the pebbles are of metamorphic origin, but some of them appear to have been derived from unchanged rocks. There are also layers of fine-grained sandstone. The prevailing color is a rusty yellow and light gray. Most of the sandstones in this country are of a rusty yellow color; No. 1, cretaceous.

9. A broad space, three hundred to four hundred feet, grassed over. The slope is complete, but it is undoubtedly made up of the sands and sandstones at the base of the cretaceous group.

10. A fine sandstone passing up into a close compact flinty rock. This is a low ridge, appearing only now and then above the grassy surface. The slope then continues down to the stream which flows through the synclinal valley about a mile wide, and then we come to the grassy slope on the mountain side inclining east again. A little below this point the creek cuts through the sandstone and black clays of No. 2, conforming perfectly to the wall of sandstone No. 1.

It is now well known that the great Rocky Mountain system is not composed of a single range, but a vast series of ranges, covering a width of six hundred to one thousand miles. There are also two kinds of ranges, one with a granitoid nucleus, with long lines of fracture, and in the aggregate possessing a specific trend; the other has a basaltic nucleus, and is composed of a series of volcanic cones or outbursts of igneous rocks, in many cases forming those saw-like ridges or sierras, as the Sierra Nevada, Sierra Madre, &c. Along the eastern portion of the Rocky Mountains, from the north line to New Mexico, the ranges with a granitoid nucleus prevail. Each one of the main ranges is sometimes split up into a number of fragments, which locally may vary somewhat from a definite direction, but the aggregate trend will be about northwest and southeast.

As I have before stated, each one of the main ranges seems to me to form a gigantic anticlinal with a principal axis of elevation, and the lower parallel ranges descending like steps to the plains, or to the synclinal valley. If, for example, we were to study carefully one of the minor mountain ranges, as the Black Hills of Dakota, or the Laramie range, where the system is very complete and regular, we should find a central granitic axis, and on each side a series of granitic ridges parallel with it, and in the aggregate trending nearly north and south. And on the eastern portion of the anticlinal, the east side of the minor ridges slopes gently down, while the west side is abrupt; and on the western portion *vice versa.* But if we take the ridges singly and examine them, we shall find in most cases that the aggregate trend is nearly northwest and southeast. The consequence is, that as we pass along under the eastern flanks of the mountain from north to south, these minor ranges or ridges present a sort of "*en échelon*" appearance; that is, they run out one after the other in the prairies, preserving the nearly north and south course of the entire system. Not unfrequently a group or several of these ridges will run out at the same time, forming a huge notch in the main range. This notch in most cases forms a vast depression with a great number of side depressions or rifts in the mountains, which give birth to a water system of greater or less extent. Such, for example, is the notch at Cache à la Poudre, Colorado City, Cañon City, on the Arkansas River, and other localities. If we were to examine the excellent topographical maps issued by the War Department, which are beyond comparison the most correct and most scientific of our Rocky Mountain region in existence, we should at once note the tendency of all the minor ranges, with a continued line of fracture and a granitic nucleus, to a southeast and northwest trend; sometimes it is nearly north and south, and then these ranges pass out or come to an end without producing any marked influence on the topography, except, perhaps, some little stream will flow down into the plain through the monoclinal rift. But when several of these minor ranges come to an end together, an abrupt jog of several miles toward the west is caused. Then frequently as the range dies out, a local anticlinal or a semi-quaquaversal dip is given to the sedimentary beds. Between the notches or breaks in the mountains, the belt of ridges or "hog-backs" becomes very narrow, sometimes even hardly visible, and sometimes entirely concealed by superficial deposits. But at these breaks the series of ridges split up and spread out so as to cover an area from half a mile to ten or fifteen miles in width. It is in these localities that the complete geological structure of the country can be studied in detail. I do not know of any portion of the West where there is so much variety displayed in the geology as within a space of ten miles square around Colorado City. Nearly all the elements of geological study revealed in the Rocky Mountains are shown on a unique scale in this locality. The same may be said, though in a less degree, of the valley of the Arkansas as it emerges from the mountains near Cañon City. I am inclined to believe that it is only in these localities that rocks older than the triassic or red beds are shown along the eastern flanks of the mountains south of Cheyenne. I have looked in vain for a single exposure of well-defined paleozoic strata from Big Thompson to Colorado City, a distance of over one hundred miles. I am now convinced that in the north, the paleozoic rocks are often concealed for long distances, although I have usually represented them by colors on a geological map by a continuous band along the mountains. That they exist continuously along the eastern margins in Colorado and New Mexico I cannot doubt, but only at these specially favored localities do they appear from beneath the triassic or red beds. They are, however, far more frequently exposed further northward, and I think much more largely developed.

Between Big and Little Thompson Creeks the ridges are very numerous and bold, and it would seem as if the massive fine-grained sandstones predominated, for they cap all the ridges, and the broken masses, often of large size, are scattered in great profusion everywhere. In one valley the abrupt side, which was composed of red sandstone, presented an unusually massive front, and in many places, are weathered into the grotesque forms so well shown southwest of Denver.

Near the head of Little Thompson the ridges are admirably well shown. Two beds of sandstone, belonging to the lower cretaceous group, seem to have broken off in the process of elevation, and so tipped over that the upper edges are past verticality. The upper cretaceous beds really form but one principal ridge, although made up of three or four subordinate ones. The sediments of these beds are so soft and yielding that they have been easily worn down smoothly or rounded off and grassed over for the most part. But by looking across it, it is not difficult to detect the black shales of No. 4, the yellow laminated chalky marl of No. 3 passing into the alternate layers of light-gray limestone and black plastic clays of No. 2. As the little streams cut through these ridges at right angles, they reveal not only the different beds, but also the dip very distinctly.

The Little Thompson begins to show evidences of enormous drift agencies in the thick deposit of gravel, the high table lands on each side of the creek, with here and there a butte with the top planed off, and over the surface is strewn a vast quantity of loose material which has been washed down from the mountains. Each one of the little streams has worn its way through the ridges of upheaval, usually making enormous gorges, but sometimes producing wide open valleys. The valley of St. Vrain Creek is one of these valleys of erosion, with broad table lands or terraces on each side, leaving the divide in the form of a continuous smooth bench, extending far down into the prairie, giving to the surface of the country a beautiful and almost artificial appearance.

The banks of the St. Vrain seem to be composed of an upper covering of yellow marl, which soon passes down into gravel. The soil appears to derive its fertility from the eroded calcareous sediments of No. 3, but it rests upon a great thickness of a recent conglomerate, cemented, in part at least, with oxide of iron. The greatest width of this valley is over ten miles, gradually sloping down to the bed of the creek from the north. The abrupt side is on the south, where a bank fifty feet high is cut by the channel of the stream. This bank increases in height toward the mountains, but becomes lower further down the stream eastward. Above this bank, southward, is a broad level plain about two miles in width, and then a gentle rise leads to another broad table plain which forms a bench or divide.

On the north side of St. Vrain Creek, near the foot of the mountains, there is a long ridge of rather rusty yellow and gray sandstone, with a trend about north 5° east, or nearly north and south. There are also two other ridges, with a dip varying between 45° and 55° east. The first ridge is about one hundred feet across the upturned edges, and there is then westward a grassy interval of three hundred feet, and then another ridge of about the same thickness, the harder layers projecting above the grassy plain from two to thirty feet. It presents the appearance, in the distance, of a high, rugged, irregular wall, or broken-down fortification, and is about three-fourths of a mile in length. These are the lower sandstones of the lignite tertiary projecting above the grassy plain.

Near the foot-hills of the mountains, about four miles south of St. Vrain's Creek, are some high cretaceous benches, extending down from the base of the mountains. They usually do not extend more than one or two miles in length before they break off, sometimes abruptly and sometimes gradually. Not unfrequently a sort of truncated cone-shaped butte is cut off from the end of some of the benches. On the summit is a considerable thickness of a recent conglomerate which has been cemented so as to form a tolerably firm rock. In this drift some fragments of the red sandstone are found, but the rocks are mostly granitic. Sometimes there is a valley scooped out between these benches and the foot of the mountains; and again, they ascend gently up to the base and lap on to the flanks. Sometimes in the interval between these benches there is a low, intermediate level or terrace about fifty feet above the valley. The higher benches are about two hundred feet above the bottom. It is to this peculiar configuration of the surface into bench and terrace, that the wonderful beauty of this region is due. In the distance southward can be seen a continuation of the ridges of tertiary sandstone as they project above the surface far in the plains, five to eight miles from the base of the mountains. There are some of these sandstone ridges from one hundred to three hundred yards apart; the intervals level and completely grassed over, so that the laminated clays or coal beds are entirely concealed from view. These ridges continue to appear above the surface now and then, nearly to Denver. Where they pass across the valleys of streams, or even dry branches, openings are made of greater or less depth and width, which give the irregular outlines to the sandstone ridges.

Between St. Vrain Creek and Left-hand Creek there is a broad plateau, about ten miles wide, which is as level to the eye as a table top. It is covered over with partially worn boulders. Near the base of the foot-hills, behind this plateau, there is a most beautiful valley scooped out, about two miles wide, which must have been the result of erosion in past times, for there is very little water in it at present.

Further southward those long narrow benches extend down into the prairie from the foot-hills. As we come from the north to the south side of the plateau, we can look across the valley of Left-hand Creek to near Boulder Valley, at least ten miles, dotted over with farm-houses, fenced fields, and irrigating ditches, upon one of the most pleasant views in the agricultural districts of Colorado. These plateaus and benches are underlaid by cretaceous clays, only here and there passing up into the yellow sandstones of No. 5, with *Inoceramus* and *Baculites.* The plateau on the north side of Left-hand Creek comes to the stream very abruptly and seems to have presented a side front to the later forces which transported the boulder drift from the mountains, the sides being covered thickly with worn rocks of all sizes. This district is very aptly called Boulder County, but the culmination of this boulder drift is to be seen in the valley of Boulder Creek.

From Left-hand Creek to Golden City the flanks of the mountains seem to be formed of the transition sandstones, or cretaceous No. 1, with all the older sedimentary rocks lying against the metamorphic rocks in such a way as to render them very obscure and the scenery quite remarkable.

Indeed, south of St. Vrain Creek the change in the appearance of the belt formed of the ridges or "hog-backs" is very marked.

As I have before stated, I believe that the agencies which produced the present configuration of the surface of the country are local and came from the direction of the mountains; and I have seen no evidence that among the later geological events there was any drift agency universal in its character as that attributed to the drift action in Canada and the Atlantic States. The forces may have acted synchronously and all over the continent west of longitude 100°, from the Arctic Ocean to the Isthmus of Darien, but the mountain ranges were the central axes from which the eroding agencies proceeded. The agency which produced the erosion and deposited the drift in the valley of a stream originated in the mountain range at the source of that stream. I shall refer to this subject from time to time, and it is one fraught with the deepest interest to the student of geology in this country, and one around which there is no small degree of obscurity. The effects are universal, however, the evidences of erosion and the worn drift materials being found on the summits of the highest ranges as well as in the lowest valleys, and each district pointing out the source of these eroding and transporting agencies in the immediate vicinity.

Since leaving St. Vrain Creek, the tertiary beds containing the coal have been aproaching nearer the mountains. North of this point the belt of cretaceous rocks has been quite wide, varying from two to five miles, but in the valley of the Boulder the belt becomes quite narrow, and forms a part of the foot-hills themselves, while Nos. 4 and 5 are entirely concealed from view.

In the Boulder Valley the tertiary coal beds are enormously developed. The Belmont or Marshall's coal and iron mines, on South Boulder Creek, are the most valuable and interesting, and reveal the largest development of the tertiary coal-bearing strata west of the Mississippi.

In the autumn of 1867 I had an opportunity of examining these mines, under the intelligent guidance of J. M. Marshall, esq., one of the owners of this tract of land, and I wrote out the results of my examinations at that time in an article in Silliman's Journal, March, 1868. In July, 1869, I made a second examination of this locality under the same auspices. The following vertical section of the beds was taken, which does not differ materially from the one hitherto published:

48. Drab clay with iron ore along the top of the ridge.
47. Sandstone.
46. Drab clay and iron ore.
45. Coal, (No. 11,) no development
44. Drab clay.
43. Sandstone, 15 to 20 feet.
42. Drab clay and iron ore.
41. Coal, (No. 10,) no development.
40. Yellowish drab clay, 4 feet.
39. Sandstone, 20 feet.
38. Drab clay full of the finest quality of iron ore, 15 feet.
37. Thin layer of sandstone.
36. Coal, (No. 9,) nearly vertical, where it has been worked, 12 feet.
35. Arenaceous clay, 2 feet.
34. Drab clay, 3 feet.
33. Sandstone, 5 feet; then a heavy seam of iron ore; then 3 feet of drab clay; then 5 feet sandstone.
32. Coal, (No. 8,) 4 feet.
31. Drab clay.
30. Sandstone, 25 to 40 feet.
29. Drab clay, 6 feet.
28. Coal, (No. 7,) 6 feet.
27. Drab clay, 5 feet.

26.	Dip, 37°.	Sandstone with a seam of clay, 12 to 18 inches, intercalated, 25 feet.
25.		Drab clay, 4 feet.
24.		Coal, (No. 6,) in two seams, 4½ feet.
23.		Drab clay, 3 to 4 feet.

22. Yellowish, fine-grained sandstone in thin loose layers, with plants, 5 to 10 feet.

21.	Dip, 80°.	Drab clay, excellent iron ore.	15 feet.
20.		Coal, (No. 5,) 7 feet.	
19.		Drab clay.	

18. Sandstone, dip. 11°. This sandstone has a reddish tinge, and is less massive than 14.
17. Drab clay. }
16. Coal, (No. 4.) } 20 feet, obscure.
15. Drab clay. }
14. Sandstone, massive, 60 feet.
13. Drab clay.
12. Sandstone.
11. Drab clay.
10. Coal, (No. 3.)
9. Drab clay.
8. Sandstone, 25 feet.
7. Drab clay.
6. Coal, (No. 2,) 8 feet.
5. Drab clay.
4. Sandstone, about 25 feet.
3. Drab, fire clay, 4 feet.
2. Coal, (No. 1,) 11 to 14 feet.
1. Sandstone.

In bed No. 23 there are three layers of sandstone, which contain a great variety of impressions of leaves. Below coal bed No. 6 there is a bed of drab clay, seven feet thick, with a coal seam at the outcrop, three feet thick; but the coal appears to give out or pass into clay as the bank is entered, so that there are ten feet of clay above coal bed No. 6.

Much of the iron ore is full of impressions of leaves in fragments, stems, grass, &c. The ore is mostly concretionary, but sometimes it is so continuous as to give the idea of a permanent bed. There are several varieties of the ore of greater or less purity. Above coal bed (5) there is a seam of iron, with oyster shells, apparently *Ostrea subtrigonalis*, or the same species found so abundantly near Brown and O'Bryan's coal mine, about twenty miles southeast of Cheyenne. Nearly a dozen openings have been made here for the coal.

These coal beds are the more valuable, and can be more easily wrought than any in Colorado. The great thickness of the coal strata has been so uplifted, and the surface worn away, that the beds are all easily accessible, and one can walk across the upturned edges of from 1,200 to 1,500 feet in thickness and then they incline eastward, and die out in the plain. I find it somewhat difficult to give a satisfactory reason why they have not been swept away or concealed by debris, as they have been in most other localities. Leaning against the sides of the mountains between South Boulder cañon and that of the main Boulder Creek, are immense walls of sandstone, possibly paleozoic or the lower beds of the trias, partially metamorphosed by heat. These walls rise to the height of 1,500 to 4,000 feet above the valley, and thus seem to have protected these formations from the erosive action, which, according to the position that I have taken in this report, is local, and must have come directly from the mountains.

A beautiful valley has been scooped out by the South Boulder, leaving a bench covered with debris between the two Boulder Creeks. Before reaching these huge sandstone walls, we pass over a portion of the cretaceous, and a great thickness of the red beds, inclining at a high angle.

Immediately south of the South Boulder Creek there is a high bench that extends up close to the base of the mountains, and is covered with drift and boulders, three miles in width, entirely concealing all the unchanged rocks. But in the valley of Coal Creek, seven beds of coal are revealed by the scooping out of this valley. These beds all incline at a high angle, about 45°, and are not easily worked. The sandstones project up above the loose material like irregular walls, and the creek itself forms a narrow passage or gorge through one of these ridges.

Between the sandstones, and apparently with very little clay either above or below, is one bed of coal four to six feet thick, which was wrought for a time, and then abandoned.

It seems to me the coal here will never be worked with profit. Above the sandstone there is another bed of coal, and above that, fire-clay; all the strata conforming and inclining between 35° and 45°. The sandstone ridge on the north side of Coal Creek becomes more nearly vertical—68°. All the beds of coal are so badly crushed together that they are rendered somewhat obscure. There are here two or three feet of clay between the layers of coal, and above the coal the clay is very irregular; sometimes thinning out entirely, so that the sandstone comes directly upon it. A large number of the sandstone ridges may be seen far out in the plains, east of the mountains, at intervals, all having the same general trend, and inclining at various angles. They rise above the grassy plains in isolated piles, like broken-down walls. These sandstones indicate the existence of coal beneath, but it would be utterly impossible to work out the sequence of these beds only at the most favorable exposures. In almost all cases the tertiary beds are so worn down and covered with superficial deposits that they are detected only in the channels of streams, or by the sandstones projecting above the grassy surface of the plains.

July 6.—With Mr. Marshall as guide, I attempted to penetrate through the sandstone beds to the metamorphic rocks up Bear Cañon, a sort of separation in the immense sandstone wall between the two Boulder Creeks. So far as I could ascertain in this cañon, the sedimentary beds lie fairly against the metamorphic rocks, and the latter incline in precisely the same direction, and at about the same angle as the former, a little north of east. There is another point that seems to me to be well shown in the range; and that is, that the metamorphic rocks are thrown up in distinct anticlinals, the same as the sedimentary beds. As soon as we pass the junction of the unchanged and changed rocks we find the granites inclining in the same direction, and a little further up there is a ridge inclining in the opposite direction, forming in the interval a valley. The angle of dip on the west side of the granitic anticlinal is 44°, a little south of west. This anticlinal feature may be local here, but I regard it as a common occurrence in the metamorphic rocks of the mountain ranges.

Here tremendous uplifts of the sandstones appear about 4,000 feet above the Boulder Valley in the plains below, and their rugged summits project far over on the granitic rocks westward, so that along the little stream immense masses have fallen down from the broken edges, a half a mile above the junction of the two kinds of rocks. I think this illustration alone furnishes sufficient evidence that the sedimentary beds once continued uninterruptedly across the area now occupied by the mountain ranges, and that these beds only form a part of what was once a gigantic anticlinal, the eastern portion of the unchanged beds remaining, while the western portion has been worn away and mingled with the debris of the plains. Further up toward the central axis of the mountain we pass ridge after ridge of granite, inclining eastward about 36°.

The process of disintegration of the rocks by exfoliation is here shown quite clearly, without regard to stratification. Immense masses of rock are weathered into rounded forms by these coatings or layers falling off. I have observed that all kinds of rocks, granites, igneous rocks, sandstones, limestones, &c., have a tendency to weather by this process of exfoliation, and the hills and mountain-peaks follow the same rule. It would seem that nature abhors sharp angles and corners, and commences at once to smooth and round them off, so that nearly all peaks and hills have this rounded appearance when closely examined. The huge masses of granite or basalt on the summits of the highest mountains are now undergoing this process of exfoliation.

The first bed of granite that lies west of the high ridge of sandstone inclines 58°, and has much the appearance of sandstone completely metamorphosed. It is of various degrees of fineness, but mostly an aggregate of coarse crystals of feldspar and quartz. There is also a bed of mica schist inclining with it at an angle of 48°. I have made use of these gneissic beds to aid me in forming a clearer idea of the true stratification or bedding of the granite, which is often obscure.

The massive beds of sandstone which form the high walls are evidently partially metamorphosed by heat. The bottom beds, which lie next to the granites, are composed of a rather coarse aggregate of crystals of feldspar and quartz, inclosing multitudes of well water-worn pebbles of all kinds, from a minute size to several inches in diameter. There are also fragments of unchanged reddish sandstone, but the inclosed pebbles are mostly metamorphic, among which quartz pebbles are conspicuous.

The inclination of the first ridge is about 33°. A portion of it is so fine and compact that it has somewhat the appearance of imperfect jasper. It varies much in texture. A most interesting feature is the separation of this inner ridge from the one just east of it. It has evidently been broken off from the summit of the next one east of it, and the whole mass carried forward westward, yet retaining nearly the same angle of inclination. This is shown by the fact that the granite rocks are thrust up under and between the ridges, showing most distinctly that this is an immense fragment of the second ridge from the inside, elevated upon the edges of the granitic rocks and carried two hundred or three hundred feet to the westward. Yet the agency that performed this movement acted so quietly that it did not disturb its position in relation to the other ridges.

The second or main ridge from the inside varies in dip from 30° to 45°. It is largely composed of pudding-stone or fine conglomerate, with layers of sandstone of various degrees of fineness. The upper beds are composed of fine-grained sandstone. The entire ridge must have had a thickness of eight hundred to twelve hundred feet.

The scenery along the flanks of the mountains at this point is wonderfully unique, and I have never known a similar example in the Rocky Mountain region. The uplift is on an unparalleled scale.

Toward the outside, or, more properly, the upper layers of this ridge become close-grained, much of it breaking into cubical blocks and forming a great accumulation of debris on the sides of the mountains. The outermost layer of this ridge, which has been worn off so as to be a low one, inclines 54°. All the beds exhibit less and less the influence of heat from the inner to the outer side, and much of the upper part is a compact, close-grained quartzose sandstone, divided into layers with smooth surfaces, and most excellent for building purposes.

The next bed is a loose red sand, so soft that the upturned edges have been worn down and completely grassed over. The upper edges of this bed are at least twelve hundred feet below the summit of the high sandstone ridge. The dip is 31°. At the foot of the slope of these red beds is a grassy valley, and then a very abrupt ascent to the edges of a thick bed of yellowish sandstone. At another locality a few yards distant a small stream, in cutting its way through this ridge, revealed alternate layers of ash-colored and yellow arenaceous clay, with some hard beds of sandstone, inclining 55°. A portion of these beds are probably jurassic. We have here an interval in the harder beds between the high sandstone ridge and the sandstones of No. 1, filled up with yielding clays and sands, which I estimated at from six hundred to seven hundred feet in thickness. Then come the sandstones of No. 1, and the gray limestones and shales of No. 2, and the chalky marls of No. 3, which are plainly visible with about the same dip. Although the grass covers the surface to such an extent that the upper cretaceous beds are not exposed, yet it is safe to suppose that the entire series of cretaceous formations, as known along the flanks of the mountains, exist here.

There is ample room, also, for a great thickness of the tertiary beds, and the evidence is quite clear that a large portion of the sandstones, clays, and doubtless beds of coal, of the tertiary period exist in the enormous plateau or table-like bench which extends down the Boulder Valley from the foot of the mountains.

The amount of loose drift material is enormous, scattered not only over the surface, but concealing to a great extent the underlying basis rocks. There is, therefore, some reason to believe that the coal may yet be found in the valley under South Boulder Creek and between it and the foot-hills of the mountains.

We find, therefore, that we have at this locality a somewhat narrow belt of the unchanged rocks, packed close together, and inclining at about the same angle, and perfectly conforming with each other, and the metamorphic rocks also. In passing up the cañon of the little stream from the Boulder Valley we cross the visible edges of cretaceous formations Nos. 3, 2, and 1, the jurassic red beds, and the paleozoic sandstones, to the metamorphic rocks. While I believe that the extensive series of coal strata all perfectly conform with the older formations, yet as we pass eastward from the Boulder Valley the dip becomes less and less until it ceases in the plains.

An important question arises as to the cause of the change in the sedimentary rocks of this region. That the sandstones forming the huge ridges have been partially metamorphosed is clear, though the traces of their sedimentary origin are as plain as ever.

The limestones of cretaceous formation No. 3 are more compact at this point than I have ever observed them northward; and the coal, along a narrow belt, is far superior to that which is found farther eastward in the plains. I am inclined to believe that the area from which first-class coal will be obtained in Colorado is very restricted, and will be comprised in a moderately narrow belt along the base of the mountains south of Boulder Creek and north of Golden City.

These changes might be attributed, wholly or in part, to the influences of igneous action in the vicinity. In the valley of the Boulder, near Valmont, there is a prominent dike of very compact basalt, which rises up like a wall, but does not seem to have disturbed the tertiary sandstones in the vicinity. Near Golden City, about twenty miles southward, close to the base of the mountains, are two large mesas, or table-lands, covered with a thick layer of basalt, which must have passed up from below in the form of a dike, and flowed over the tertiary rocks.

These are the only instances of eruptive rocks observed by me from near the South Pass to the Arkansas, a distance of nearly four hundred miles. In the Middle Park, just west of Long's Peak, and in the

South Park also, are numerous examples of the outpouring of igneous material. That internal heat connected with these igneous outbursts may have affected the sedimentary rocks in the Boulder district, and rendered the coal more compact and anthracitic, under pressure, seems to me possible, at least. The rocks which appear to have been affected by heat are seen only for a few miles south of the Boulder—from five to ten miles. South of that no effects whatever have been observed.

The next finest exhibition of coal in Colorado to Marshall's mine is that of the Murphy mine, on Ralston Creek, five miles north of Golden City. The coal bed is nearly vertical in position, and varies in thickness from fourteen to eighteen feet, averaging sixteen feet from side to side. There are nine feet of remarkably good fire-clay on each side of the coal, and above and below, or on the west and east sides, are the usual beds of sandstone. This mine is very near the foot of the mountains, and the belt of sedimentary rocks, which are all nearly vertical, is very narrow here—not more than half a mile in width—and are mostly concealed by debris.

Mr. Murphy thinks that there are eleven beds of coal within the distance of one-fourth of a mile, all nearly or quite vertical in position, of which the one opened is probably the oldest. The mine is opened on the north side of the creek, and may doubtless be followed above water line several miles to the northward, toward Coal Creek.

On the south side of Ralston Creek the same bed has been opened, and the indications are that it may be followed the same way southward toward Golden City. The entire surface is so covered with superficial deposits, and grassed over, that it is impossible to work out these beds in detail, and the artificial excavations afford us the most reliable knowledge. A hundred yards or more west of the coal bed there is a high ridge running parallel with the mountain range, capped with lower cretaceous sandstones No. 1.

This ridge extends southward, with some interruptions, beyond Golden City.

At Golden City the upheaved sedimentary rocks are so swept away that the metamorphic foot-hills are plainly visible. No rocks older than the red beds or trias are exposed, and these somewhat obscurely. The red and gray sandstones lie close on the sides of the metamorphic rocks, inclining 30° and 54°. In the trias there is a bed of silica or an aggregation of very fine grains of quartz which has attracted some attention, and close to it a layer of bastard limestone or calcareous sandstone. All the beds dip at a high angle and lie side by side, so that one can walk across the upturned edges of them all, from the metamorphic to the summit of the tertiary. Outside of the cretaceous beds there is a small valley of erosion, and then come the tertiary beds. The strike of the coal strata is very nearly north and south, and, so far as I could ascertain, the sequence of the beds from within, outward, is as follows:

1st. Rusty, yellow, soft sandstone. 2d. A bed of fire-clay. 3d. Coal about eight feet thick. 4th. Fire-clay. 5th. Rusty, yellow sandstone.

The clay underneath the coal appears to be ten or fifteen feet thick, with one or two unimportant seams of coal. These beds have been so elevated that the upper edges have passed verticality 5° to 10°. The clay is much used for fire-brick and potter's ware. In the bed of sandstone, above the coal, we found several impressions of leaves of deciduous trees, among them a *Platanus*, probably *P. haydeni*. From these we pass across the edges of a series of beds of sandstone, with intervening strata of iron ore. The thickness of all the tertiary beds here must be 1,200 to 1,500 feet. Near the outside is a bed of pudding-stone, and outside or above this, the bed of potter's clay, which supplies the pottery at Golden City. About midway in this series of beds an entrance has been made exposing a second bed of coal. The surface is so grassed over that it is quite impossible to make out the full series of beds clearly, but the softer strata are well shown by the depressions between the beds of sandstones.

The north mesa is two and one-half miles long and about one mile wide. The south one is four miles long and about a mile wide. This one has an irregular surface and gradually slopes down eastward until it becomes a low ridge of tertiary sandstones and clays. The wall of basalt that surrounds the top is nearly perpendicular most of the way round, from fifty to one hundred and fifty feet in height. The lower portion of the basaltic bed on the north side of the south mesa is very vesicular, full of rounded porous masses somewhat like slag, and rests upon the slightly irregular surface of a bed of fine fire-clay, which contains traces of vegetable remains. Below the fire-clay are alternate beds of sandstone and arenaceous clay, inclining slightly east, and evidently protected from erosion by the hard cap of basalt. These beds are plainly tertiary lignite, and must be six hundred to eight hundred feet thick. The lowest bed of vesicular basalt is evidently more recent than the columnar bed above.

Golden City is a thriving little town, located near the embouchure of Clear Creek from the mountains, which is called the "Golden Gate." Clear Creek Valley is very fertile, and, in looking down upon it from the top of the mesa, it appears like a finely cultivated garden. The ridges of upheaval or "hog-backs" near Golden City are small and unimportant, owing to the erosion which has worn them down. But proceeding southward a short distance they increase in size. The tertiary ridges are most conspicuous until we reach Mount Vernon, about five miles south of Golden City, where the older formations are largely displayed. Here the tertiary beds are tipped past a vertical position and seem to incline toward the mountains; but this is more apparent than real; the top portion leaning over, while deeper in the earth the strata incline at a high angle from the mountains.

Green Mountain is a lofty, grass-covered hill, and is entirely composed of the coal strata, while to the west of it is a nearly vertical ridge of sandstone. Just inside of this ridge, or beneath it, is a coal bed which has been opened by Mr. John A. Roe. The entrance to this mine is the finest I have seen in Colorado, and is 170 feet in length, through 141 feet of sandstone with a slope of 45°. The sides and roof of the entrance are not protected. The bed of coal is nearly vertical in position at this point, though at some places where it is not wrought it inclines east 70°. There are three seams of coal, 4 feet each, in thickness with 3½ feet of clay intervening. Below the coal there is a bed of clay 5 feet thick, and above 3½ feet arenaceous clay. The coal is close, compact, and makes an excellent fuel, and Mr. Roe, who is an old Pennsylvania miner, considers it better than the bituminous coals for all domestic purposes, but for generating steam and smelting ores he regards it as inferior. The ash is white, resembling pine-wood ashes, and the quantity is small. The coal at Murphy's, on Ralston Creek and Golden City, leaves a red ash. There are no cinders, and in burning it gives a bright, clear flame; and although it burns iron, it does not give sufficient heat to weld it. I believe this to be a continuation southward of the Golden City bed. It is also the lowest of the coal strata in this region, for in the valley immediately west and on the sides of the ridge can be seen the dark clays of the cretaceous beds. This ridge is very high at this place, and is composed of the sandstones of No. 1, and a portion of the red beds or triassic (?). Still further west are two or three rather low ridges of yellowish-gray and red sandstones, which cover the gneissoid rocks of the foot-hills of the mountains. By far the largest ridge here is the one containing the sandstones of No. 1, but it soon splits up into smaller ridges in its southern extension.

About four miles further south, in the cañon of Bear Creek and Turkey Creek, there are fine exhibitions of the beds of upheaval. The chalky shales of No. 3, with abundant specimens of *Inoceramus problematicus* and *Ostrea congesta*, form a low rounded ridge; then comes a narrow valley worn into the black shales of No. 2; and then a high ridge of massive sandstone—No. 1—inclining 30° to 35°. On the western side of this ridge we see the projecting edges of the sandstone capping the ridge, and underneath the variegated marls and sandstones, with some of the brick-red beds. Then comes a series of rather low, rugged ridges; first a layer of sandstone and loose brick-red sand with gypsum; dip 29°. Second ridge, a light gray sandstone with a rusty, yellowish tinge; dip 34°. Then come three or four small ridges of deeper brick-red, or almost purplish-red sandstone; dip 29°. The intervals between these ridges are composed of arenaceous shale. Among the red sandstones are two thin layers of bluish limestone, which is burned into lime.

The foot-hills of the mountains are composed of gneissoid rocks. They form a wide belt or range below the main or Snowy Range, rising 1,500 to 2,000 feet above the unchanged rocks. These metamorphic ridges or hills are well grassed over in many instances, and rounded, and so covered with debris that it is almost impossible to see the layers in position.

On the little creek there is a small mill for grinding the gypsum into plaster for various economical purposes, and also for sawing the sandstone into forms for architectural purposes. The gypsum is amorphous, but very white and pure, and would make the finest of casts and moldings. Some of the layers are susceptible of a high polish like the California marbles, only they are of a more uniform white color.

Up among the foot-hills, good crops are raised, especially all kinds of garden vegetables. As fine wheat as I have ever seen was growing on Mr. Morisson's farm, at an elevation of at least one thousand to fifteen hundred feet above Denver.

At Harriman's, on Turkey Creek, is an excellent place to observe the junction of the sandstones and the gneissoid rocks, and I could not determine that there was any discordance, the dip of all being 25° to 35°. The slopes of the hills, as well as the rocks themselves, show the inclination very clearly. The metamorphic rocks are distinctly stratified as any sandstones, and we find alternate beds of syenite, mica schist, hornblende slate, coarse aggregated quartz, feldspar, and mica, regular gneissoid rocks, inclining at a high angle in the same direction as the sandstones.

For a long distance there is an apparent conformability of the sedimentary rocks to the metamorphic; but I am inclined to think that it is not real or permanent. Both north and south of this point the two classes of rocks do not conform.

Near the summit of the sandstone ridge No. 1, on Turkey Creek, there is an asphaltum spring, which has been wrought for oil. A considerable thickness of the sandstone seems to be thoroughly saturated with the pitch or bitumen, and between the layers of the sandstone are accumulations of the tar. This spring is located on the east side and near the summit of the "hog-back."

About twelve miles southwest of Denver, between Turkey and Bear Creeks, are some remarkable soda lakes, which are of unusual interest. They are the property of Dr. Burdsall, of Denver, in whose company I made as careful an examination of them as my time would permit. There are four of these little lakes, and are all located on middle cretaceous rocks. The principal one lies just east of a low rounded ridge of cretaceous shale, No. 3, and is surrounded on the other sides by low ridges of superficial sand and gravel. A little west of this cretaceous ridge there is a lake, a fourth of a mile in length, but on account of the springs flowing into it from the sloping sides of the sandstone ridge No. 1 the water is not strong. The black shales of No. 2, cretaceous, underlie this lake. The soil for twenty feet in depth is fully impregnated with the soda; and on the surface of one of the lakes is a crust which looks like dirty ice. A shallow ditch which Dr. Burdsall has made out into the lake a few feet, has a deposit of sulphate of soda at the bottom in a partially crystalline state, one and a half inches thick. Three and a half barrels of the water make one barrel of the sulphate of soda, and three pounds of the soil, well leached, makes one pound of the salts. The salt, by analysis, contains sixty-three per cent. of the soda, and the water about thirty-three per cent. It contains carbonate of soda, sulphate of soda, chloride of sodium, sulphide of calcium, and a trace of magnesia. It would seem that these deposits of soda must at no distant period play an important part in the industrial operations of Colorado. These soda salts can be manufactured into bicarbonate of soda, can be used in refining gold and silver, also for the manufacture of glass with silicic acid. There is an unlimited amount of soda at this locality, and it can be procured at a mere nominal cost.

Within a few yards of these lakes, and located in the black, shaly clays of cretaceous formation, No. 2, are considerable quantities of brown iron ore of superior quality—as good as the best observed in the boulder coal strata. It occurs in the form of concretions, and occupies a very limited area.

CHAPTER II.

FROM DENVER TO COLORADO CITY.

The city of Denver is located on the tertiary rocks which contain the coal beds of the west, about ten to fifteen miles from the base of the mountains. The surface is so thickly covered with superficial drift deposits that the basis rocks are seldom seen; but we have every reason to suppose that the same beds of coal that are exposed by the uplifting of the formations along the immediate flanks of the mountains, extend eastward into the plains, and of course underlie, at certain depths, the city of Denver.

As we pass southward, up the valley of the South Platte, we find the tertiary sandstones exposed occasionally in the banks of the river; and near the cañon a seam of coal has been opened and worked to some extent. The tertiary beds extend quite close up to the foothills of the mountains, leaving a comparatively narrow space for the exhibition of the older, unchanged rocks. Still, we may walk across the upturned edges of them all and study them with care.

The valley of the South Platte presents a fine display of the terraces; and the drift, filled with water-worn rocks, is very thick. The sandstones of the tertiary formation are also plainly seen, appearing to be nearly horizontal, although not more than ten miles in a straight line from the metamorphic rocks. The whole prairie country has been so planed off that it is finely and gently rolling, and the drainage is excellent. The streams which flow from the sides of the mountains are fed by perpetual springs, and are consequently persistent and uniform in their amount of water, affording the best water-power in the country.

From the soda lakes to the great "divide" the cretaceous and tertiary beds, outside of the No. 1 sandstone ridge, are smoothed down and grassed over so that they are not conspicuous, though there are exposures enough to guide the geologist. They are so concealed by superficial gravel and sand that they present no good sections either to show the strata or dip. This regularity of the surface renders the Platte Valley, as well as those of its branches, remarkably fine for farming and grazing, and vast herds of cattle already cover the grassy hills and plains. The terraces and benches which extend down from the foot of the mountains are well shown.

Along the Platte River, near the cañon, a coal bed was opened at one time, but now it is covered with loose material which has fallen from above, so that it is entirely concealed. The strata here are nearly vertical. There are two beds of coal, in all about five feet thick, separated by about two feet of clay. The coal is not very good, and has not been used for three years. It is probably the same bed seen at Golden City, thinning out southward.

Along the Platte and Plum Creeks, the streams cut heavy beds of boulder gravel and fine sand, and it is under this deposit the coal is found. The valleys of the South Platte, and its branches, between Denver and the mountains, are exceedingly fertile and productive, and at this time they are covered with splendid crops. Nearly or quite all of the available bottom lands are already taken up by actual settlers, and are under cultivation. The present season has been unusually favorable for farming throughout the west.

The plain country south of Denver comes close up to the foot of the mountains, so that the belt of upheaved sedimentary rocks grows narrower and narrower until, a few miles south of the Platte cañon, they cease entirely for a time. The ridges are very high, ranging from four hundred to six hundred feet above the bed of the Platte. To the southwest can be seen, rising like a range of mountains, the high "divide" between the waters of the South Platte and Arkansas Rivers, covered quite thickly with pines.

The first main ridge contains a few layers of No. 2; alternations of clay and sand passing down into the sandstones of No. 1. This ridge is quite massive and inclines 43°. In the channel of the South Platte, the distance from the outside of the ridge containing the sandstones of No. 1 to the metamorphic rocks, is not more than half a mile. From this point to the "divide" the ridges are split up and much crowded. The reddish and variegated sands are worn, by atmospheric agencies, into the most wonderful and unique forms, equal to the "Garden of the Gods," only on a much smaller scale. Here also the red and variegated sandstones jut up against the metamorphic rocks as if the continuity was unbroken. Indeed, the apparent conformity is complete.

The hills of the first range, composed of metamorphic rocks, are curiously rounded and grassed over, and are made up of a reddish, decomposing granite. But, as we ascend, these peaks or rounded cones become sharper, the sides more rugged, and the rocks more compact.

As we go southward the indications of beds of jurassic age become more and more feeble. Under the massive sandstones of No. 1 are a series of yellow and white sands and sandstones passing down into brick-red sands. Among this series of variegated beds are two thin beds of limestone. One of these is a very white rock, and on its weathered surface are small masses of chert, which appear to have the structure of corals. This bed is six or eight feet thick. Separated by eight or ten feet of sandstones is another layer of bluish limestone, which is much used for lime. I have never been able to detect any well-defined organic remains in these beds, but I believe a portion of them, between the lower cretaceous No. 1 and the true red beds, are of jurassic age; and it is even possible that a portion of the red beds are of that epoch.

From the point where the Union Pacific railroad crosses the Laramie Mountains to Colorado City, I have been unable to find any well-marked carboniferous or silurian rocks. The red sandstones, which I have been accustomed to regard as triassic, jut up against the metamorphic rocks, or are the only exposures that meet the eye of the geologist. I do not believe that the carboniferous beds are altogether absent, for limestones of considerable thickness, and containing characteristic fossils, occur at Granite Cañon, on the Pacific railroad, high up on the margins of the mountains; and also at Colorado City, about two hundred miles to the south. In this long interval I have been unable to discover any well-defined carboniferous or silurian rocks, yet I am inclined to think that the carboniferous beds, at least, exist underneath all the other sedimentary rocks, but are not exposed by the upheaval.

About five miles south of the Platte Cañon the upheaved ridges come close up to the mountains, and are not worn away, but form the northern side of the divide, so that the entire series of unchanged rocks known in this region are exposed in regular continuity. A little further south we come to a series of variegated beds of sands and arenaceous clays, nearly horizontal, resting on the upturned edges of the older rocks. These beds form the northern edge of an extensive tertiary basin of comparatively modern date, either late miocene or pliocene age. From the point of their first appearance, about five miles south of the South Platte Cañon to a point about five miles north of Colorado City, these beds jut up against the foot-hills of the mountains, inclining at a small angle, never more than five to eight degrees, and entirely concealing all the older sedimentary rocks. The upheaved ridge entirely disappears. Far off to the eastward stretches this high tertiary divide, giving rise to a large number of streams, as Cherry Creek, Running Water, Kiowa, Bijou, and other creeks. Through this basin also flows Monument Creek, which has become so celebrated for its unique scenery. The beds of this formation are of various colors—reddish, yellow, and white—and of various degrees of texture, from coarse pudding-stones to very fine-grained sands or sandstones. There is very little lime in the entire series of beds. There is much ferruginous matter in all the beds, to some of which it gives a rusty brown color. The valley of Plum Creek is scooped out of this basin. The high ridge to the eastward is capped with coarse sandstones and pudding-stones. Along the immediate sides of the mountains the rocks are mostly coarse pudding-stones, the water-worn pebbles varying in size from a grain of quartz to a mass several inches in diameter. But as we recede from the mountains, eastward, the sediments become finer and finer until the coarse pudding-stones disappear. I am of the opinion that the materials composing the beds of this group have been derived from the mountain ranges and vicinity. In their general appearance the rocks of this group resemble the prevailing rocks which cover the country from Fort Bridger to Weber Cañon, and also a series of sands and sandstones along the Gallisteo Creek below Santa

Fé, which I shall call the Gallisteo sand group. To this group of modern tertiary deposits I have given the provisional name of the Monument Creek group, and they occupy a space of about forty miles in width from east to west, and fifty miles in length north and south.

Continuing our course southward, we find some curious mesas in the valley of West Plum Creek. We ascended one lofty butte, with a flat table summit, situated west of the Plum Creek road. The top of this butte is about one thousand feet above the road, and is capped with a rather close-grained, cream-colored rock, which looks quite porphyritic, fifty to one hundred feet thick, and plainly of igneous origin. It fractures into slabs which have a clinking sound. The beds below are quite variegated, of almost every color and texture, mostly fine sand, brick red, deep yellow, rusty red, white-ash colored, dull black, &c. The rusty iron layers sometimes form a sort of limonite, but are composed largely of an aggregate of water-worn pebbles cemented with the silicate of iron. There are also thick beds of quartzose sandstone, or an aggregate of crystals of quartz and feldspar, so compact as to look like a coarse granite. These large masses afford good illustrations of the process of weathering by exfoliation.

The evidence is clear in a number of localities that at a late period in geological history there were dikes or protrusions of igneous material which flowed over these Monument Creek sandstones in broad sheets or beds; and these broad, table-top buttes and mesas are the evidences that are now left after erosion.

This modern tertiary basin is very interesting as the introduction of a new feature in the geology of this region. The appearance of the country also undergoes a decided improvement. The great divide is covered rather thickly with pine timber. It is full of excellent springs and fertile valleys which give origin to numerous streams. The grass is excellent and abundant, even upon the summits of the table lands. For a distance of ten miles about the sources of Plum Creek the red beds or triassic jut square against the sides of the metamorphic foot-hills of the mountains. The projecting summits of the upturned ridges gradually fade out in importance. They have also lost their usual regularity, and are split up into an indefinite number of fragments of ridges, varying in dip from ten to forty-five degrees. Near the water divide these ridges gradually close up again toward the foot of the mountains and are entirely concealed by the sands and arenaceous clays of the Monument Creek group.

In the valley of West Plum Creek and its branches, as they emerge from the mountains, we have a fine exposure of the sedimentary beds. The coarse, yellowish-gray sandstones and pudding-stones of the Monument Creek group incline slightly, perhaps three to five degrees. Then come the sandstones of the lignite tertiary, inclining twenty-five degrees. Then west of West Plum Creek are some ridges of cretaceous rocks. The first ridge is made up of a rather impure limestone, filled with well-defined species of Inoceramus and other shells, of No. 3 or middle cretaceous. The next ridge west is composed of No. 1, and the intermediate valley is underlaid with the shales of No. 2. Among the brick-red ridges is one low ridge composed almost entirely of gypsum—an unusual development of this material—to the thickness of thirty or forty feet.

There is an extensive series of low ridges of red and gray sandstones extending up the base of the mountains.

The high portion of country, which is plainly visible from Denver when looking southward, and from the Arkansas River looking northward, would seem to have been protected from erosion by causes which I cannot yet well explain. The water divide is the long bench which extends down from the very base of the mountains eastward, and forms the line of separation between the sources of the streams which flow southward into the Arkansas on the one side and into the South Platte on the other. This water divide is well worthy of especial notice, inasmuch as it is composed of the Monument Creek formation, and juts up against the almost vertical metamorphic rocks, retaining its nearly horizontal position, and perfectly concealing all the older rocks for at least five miles north of the line of separation.

The valleys of Plum Creek and of its branches are quite wide, and are scooped out of the modern deposits so as to form a most beautiful and fertile lands, while on each side a bench extends down from the mountains like a lawn. The series of older rocks are exposed by the stripping off of the newer tertiaries in the valley of Plum Creek. The bench on the north side conceals them, for the most part, close up to the foot of the mountains, while on the south side they are entirely concealed until they reappear near Colorado City.

The divide forms a high ridge with a mesa-like top, stretching far eastward beyond the horizon, covered with pines. On each side the beds of whitish-yellow and reddish sandstones appear like fortifications, holding a nearly horizontal position. Near the foot-hills there is a narrow valley, perhaps one-fourth of a mile wide, and lying against the sides of the mountains are remnants left after the erosion. I at first mistook them for the red triassic beds, but on a close examination I found them to be a coarse aggregate of feldspar and quartz, colored extensively with iron. There are inclosed in the rock various water-worn pebbles of all sizes and textures. This rock decomposes readily, especially by the process of exfoliation. The whole rock is so massive and compact that it might easily be mistaken for a metamorphic sandstone.

Just south of the first branch of Monument Creek there is a fine exhibition of the erosion of the sandstones. At one locality they lie snug up against gneissoid rocks, showing the discordant relations perfectly. These illustrations seem to show plainly that the sediments of this recent tertiary deposit have all been derived from the disintegration or erosion of the metamorphic rocks and perhaps the older sedimentary beds in the immediate vicinity.

In a beautiful little basin near Monument Creek, which leads to the creek, is a lone pillar or column of sandstone, three-cornered, with the strata perfectly horizontal, about thirty feet high. The sands composing this are coarse and of a yellowish or whitish color. It has been for a long time a favorite object for the photographer.

At one point on Monument Creek the red granites, high up on the mountain side, show the perpendicular lines of cleavage in a marked manner. Some of the openings are several feet wide. The strike of these lines of cleavage is about southwest and northeast.

For a considerable distance, some ten or fifteen miles, along the immediate base of the mountains, on the west side of Monument Creek, the long, smooth, grassy benches slope down toward the creek, sliced as it were or cut by the numerous little branches. These lawn-like slopes or benches vary in height. Sometimes on the side of a little branch, where the valley is deep, there is an intermediate terrace or step to the higher ridge.

All these valleys seem to be occupied by farmers and stock-raisers. Almost every available spot is taken up by actual settlers.

The first range of mountains on the east side, from the divide to a point near Colorado City, appears to me to present a fine illustration of what I have called an abrupt anticlinal; that is, only the abrupt side of the western slope appears here. The eastern side has either been worn away or was never elevated to a great height, and is now concealed by the recent deposits. The summit of the metamorphic ridge projects far over the base of the mountains, and the western side of the monoclinal shows a gentle slope. That this eastern portion of the metamorphic anticlinal may have been elevated and then fell back, or may not have been elevated at all and still exists beneath, is shown from the fact that the sedimentary ridges or "hog-backs" gradually diminish in dip to the point of concealment.

The little streams which flow into Monument Creek, as well as the creek itself, cut through a coarse material of various colors with irregular layers of deposition. Sometimes a layer is hardened into a coarse sandstone, and then comes a thin layer of ironstone or impure limonite, but the whole is a quartzose material and rather coarse. There are now and then thin seams of fine sand or clay. Near the stage station there is a bluff of rather massive whitish sandstone, with some thin beds of clay at intervals. There is much iron in these rocks, and this aggregates in the form of a rusty layer, quite hard. The light-colored sandstones below are weathered into most singular columnar or monument-like forms, with this layer of rusty sandstone as a cap protecting the summits. There are some dark bands of arenaceous clay, and in the sandstone a few rounded concretions.

About six miles north of Colorado City the upheaved ridges or "hog-backs" reappear from beneath the quartzose sandstones of the Monument Creek group. The white massive sandstones of the lower cretaceous lie high on the mountain side. The first ridge that we pass through along the road is a whitish brown, rather yielding sandstone, with rusty yellow portions, with very irregular laminæ of deposition. The strike is southwest and northeast, and the dip 32°. This is a bed of the lignite tertiary.

High up on the sides of the mountains, for ten miles or more about the Soda Springs, there is a great thickness of red porphyritic granite, inclining from the mountains in well-defined ridges, like sandstone. From their very deep rusty-red color, I regarded them as sandstones until I made a close examination of them. They have a well-marked dip of forty-five to fifty degrees, somewhat less than the massive granite rocks which form the nucleus. All these ridges rise like steps toward the range of which Pike's Peak forms a part, with the sloping sides toward the northeast and the summits leaning over toward the axis of elevation.

These very red granitoid rocks have formed a very conspicuous feature on the eastern side of the mountains for thirty miles or more north of Colorado City; and, as they readily decompose, the hills and roads are paved with the crystals of feldspar and quartz. The constituent which predominates is feldspar, which gives the red color. This rock is composed of a coarse aggregate of quartz and feldspar with a little black mica, and now and then a little pencil-like crystal of hornblende. The rock itself does not seem to be so red, but the debris has a dull rusty-red color in the distance. Upon the summits of the mountains about Pike's Peak are columns of massive granite—immense rounded masses, standing one upon the other, giving a most picturesque appearance to the scenery, and affording fine illustrations of the style of weathering.

The unchanged rocks are here seen resting directly upon these dull reddish granites. The lower beds are composed of a more or less fine-grained sandstone, with some small pebbles, variegated in color, passing up into rocks of a semi-crystalline texture. Most of the rocks appear as if they had been partially changed by heat. There is every variety of texture, mostly silicious, but some layers appear to be an impure limestone.

For a space of about ten miles from north to south, and an average width of five miles from east to west, about Colorado City, all the unchanged rocks are displayed in a unique and remarkably clear manner. The ridges of upheaval are spread out over an unusually wide space. Here every formation known in this region is distinctly revealed to the scrutiny of the geologist.

Beginning in the plain country we have the sands and sandstones of the Monument Creek group in a perfectly horizontal position, and separated from the older rocks by a valley about half a mile wide. It is through this valley, which runs nearly north and south, that the road passes. The Monument Creek group is seen on the east in the form of a rounded grassy range of hills; while on the west side the cretaceous formations are exposed in the form of upheaved ridges. I have no doubt but that this intervening valley is underlaid by lignite tertiary beds, for as we enter it from Monument Creek valley we have an exposure of the sandstones of this group for a little distance, revealed by the stripping off of the Monument Creek sands by erosion. They very soon pass beneath the more recent deposits. On the west side of the road, near Camp Creek, which flows through what is called the second "Garden of the Gods," we find the chalky shales of No. 3 with *Inoceramus* and *Ostrea congesta* in great abundance. All the cretaceous rocks, including the massive sandstones of No. 1, are finely displayed in this region, and No. 1 forms a most picturesque and nearly vertical wall for six to ten miles, as it were inclosing the "Garden of the Gods." There is one peculiar feature presented by these nearly perpendicular walls of sandstone, and that is, two quite distinct lines of cleavage, but not quite as regular or as well defined as in the gneissoid rocks of the mining regions. These lines cross each other, one set with a direction northwest and southeast, and the other southwest and northeast.

The rocks included in this wall-like ridge are layers of fine black shale, fine sandstone with bits of vegetable matter, and a thin seam of earthy lignite. Then come beds of whitish sandstones, with thin layers of limestone made up of indistinct fragments of fossil shells, with bed of snowy gypsum; then a series of whitish, yellow, and brick-red sandstones, with intervals of loose, laminated sands, which form a kind of grassy valleys. In passing up the Fountain Creek valley we cross the upheaved edges of twenty or thirty of these fragmentary ridges, all inclining at various angles, from ten degrees to sixty degrees. It is to the peculiar weathering of these variegated upturned ridges of sandstone that the wonderfully unique scenery of the "Garden of the Gods" is due. In some localities some of these beds seem to pass over beyond verticality 3° to 5°. The composition of these sandstones is mostly fine sand, but often it is an aggregate of minute particles of quartz, with some small, rounded pebbles. All the beds exhibit the indications of ripple marks, irregular lines of deposition, and in most, the water-worn pebbles are small, but sometimes they are from six to ten inches in diameter. The upper portions of the variegated beds are a light brick red, with spots and irregular layers of whitish sandstone.

As we pass to older beds this red color deepens until it becomes a dull purple hue. There are in all these sandstones a great many irregular seams of gypsum. Everywhere among these curious projecting ridges of sandstone are beautiful grassy intervals. To show the irregularity of the dip of these rocks, the ridges that give the most marked features to the picturesque scenery incline eighty to eighty-five degrees, and then immediately west are several low ridges dipping fifteen to twenty degrees.

There is a somewhat extensive cave in the north portion of the sandstone ridge that forms the entrance to the "Garden of the Gods." It is caused by the washing away of a soft layer, about three feet thick, by a little stream of water that trickles down from the summit of the ridge. These vertical ridges of red sandstone rise above the surface about two hundred and fifty feet. Just east of the entrance or gate, about fifty yards, is a wall of white sandstone, with seams of impure, gritty gypsum running through it in every direction, forming a kind of net-work. The strike of these ridges is nearly north and south.

At Crater's Falls, above the soda springs on Fountain Creek, there is a remarkable cañon, in which the unchanged sedimentary rocks are seen to rest directly on the red porphyritic granites. At no point along the eastern base of the mountains, from Laramie Peak southward, have I seen the two classes of rocks so fairly in apposition. The metamorphic rocks beneath are quite massive—a deep rusty red; an aggregate of crystals of feldspar and quartz, with some black mica. The cleavage lines are shown with great distinctness, but the lines of stratification in the two kinds of rocks do not precisely correspond. I think that the strata of both groups incline in the same direction, but the granites seem to be more steeply inclined. As I have before remarked, there seems to be a conformity in very many localities, and sometimes extending over large districts, between the unchanged and changed rocks, but I am inclined to regard this conformity as more apparent than real.

The rock which rests directly upon the granites at this locality is a sandstone, totally unchanged, as if it had been deposited on them in cool and rather quiet waters. It is composed of minute crystals of quartz, considerably rounded by attrition, and cemented with silicate of iron. This sandstone is quite massive, with streaks or seams of small pebbles. We have them resting upon the granites, then alternate layers of light gray, and rusty reddish sandstone—forty feet; then a very deep dull purplish sandstone with dark spots—two hundred feet. Above this a thinly laminated yellowish-white limestone, of various degrees of fineness, with vast quantities of crinoidal remains, some corals, small univalves &c. This limestone must be from three hundred to four hundred feet thick. The dip of the rocks is distinct, as the little streams have cut the most perfect sections. Sometimes masses of these rocks are lifted high on the summits of the mountains, in an almost horizontal position, then again they dip ten, twenty, or thirty degrees in different directions.

A few hundred yards to the northeast of the Crater Falls, on Fountain Creek, there is a little branch which flows down from the mountains, and has cut out of the rocks a most remarkable cañon. The limestones and sandstones are here shown most perfectly in the vertical walls, for a mile or more resting on the granites below, and inclining not more than 5° to 10°.

About four miles northwest of Colorado City is what is called the second "Garden of the Gods," through which flows Camp Creek. The area is much smaller than that of the first Garden of the Gods, but the scenery is even more remarkable. The entrance is through a kind of gateway, cut by the creek at right angles to the ridge of lower cretaceous sandstone No. 1. This ridge forms high walls, with a dip to the east of 55° to 60°. Then comes, inside of this wall, a narrow belt of what must be jurassic limestone, some portions being of a bluish color and brittle, filled with indistinct animal remains. Then comes the gypsiferous sandstone, with a bed of snowy gypsum, gradually passing into light brick-red, and deep, dull, purplish sandstones. Here again the sandstones are worn into wonderful shapes—columns, peaks, &c. All the sedimentary rocks are reduced to a narrow belt, and the ridges are crowded together into a space of hardly a mile in width, and on the foot-hills of the mountains are the deep, dull, red sandstones and limestones of the carboniferous resting upon the red granites. The walls of the Camp Creek cañon show all the carboniferous beds in their relation with the granites most perfectly. Upon the weathered surface of the reddish limestones I found a number of specimens of brachiopodous shells.

A short distance north of this cañon, the jurassic and carboniferous beds are seen in a nearly vertical position, and lying in perfect apposition, showing complete continuity. It is therefore my opinion that there is no discordancy in the unchanged beds, from the granites up to the Monument Creek group. The latter never conform to the beds below, while I am inclined to regard all the instances of apparent conformity of the sedimentary rocks with the metamorphic as not real but accidental.

As the ridges emerge from beneath the Monument Creek group at the north end of the second Garden of the Gods, the trend is a little east of south, and they finally bend around so that they jut up against the base of the mountains a little way south of Colorado City, with a trend nearly east and west.

About five miles east of the base of the mountains, and four miles northeast of Colorado City, Mr. Gehrung has a land claim where a coal bed crops out of the bank of a creek. Above the coal is about eight or ten feet of clay, and below there is also a bed of clay, and the coal above and below gradually passes into the clay. This clay is filled with fragments of vegetable matter, some seeds and plants. The clay passes up into fine sand. In the distant hills, the beds of whitish massive sandstones are weathered into fortification-like bluffs. The coal is very light, varies much in thickness, from a few inches to five or six feet, and seems to be a sort of jet. There are several other localities where the carbonaceous clay crops out in the valleys of the little branches, and it occurs in the Monument Creek group, and therefore must be of very modern date. There are also, in the clays above and below the coal, considerable quantities of impure brown iron ore.

Perhaps the feature of the greatest general interest in this region is the Soda Springs, which are located about three miles above Colorado City, in the valley of Fountain Creek. The water issues from the ground very near the junction of the sedimentary and metamorphic rocks, close by the base of Pike's Peak. The scenery around them is grand beyond any that I have ever seen in the vicinity of any other medicinal springs.

There are four of them. The first one is close to the road and within fifty feet of the creek, and perhaps at this time ten or fifteen feet above its bed. The violent bubbling up of the water would indicate the issue of a large supply, but there can hardly be a gallon a minute. For a distance of sixty feet or more around the spring there is a deposit or incrustation in thin layers. Its thickness I could not determine, though it is probably not more than six or eight feet. About twenty-five feet west of the present opening there is another which formerly gave exit to the water. It is about five inches in diameter. The sediments deposited around these springs seem to be filled up with foreign matter, introduced during deposition. Portions of the deposit are very hard and filled with small cavities, lined with a whitish, partially crystalline material, probably carbonate of lime or gypsum.

About one hundred yards above the first spring is the second one, on the right side of the creek. This is much the largest one and has formed a basin six or eight feet across, from the center of which boils up a most violent current, so that one would suppose there was water enough to make a good-sized trout brook, and yet not more than five or six gallons a minute issue from it. A small stream about four inches wide, and an inch deep, passes off into the creek. About this spring, also, there is a large deposit, which is rounded off on the side toward the creek by the overflow of the water from the spring.

On the opposite side of the creek, not more than twenty feet from it, and located about ten feet above it, is a third small spring. The water is stronger than that of the others and is used principally for drinking purposes. The cavity in this deposit is about twelve inches in diameter and the water twelve inches deep, and the bubbles rise continually and energetically, but not more than half a gallon of water a minute passes off. There is now a constant deposition of a whitish substance from the spring, and it extends to the margin of the creek. Between the second and third springs are two massive red felspathic granite boulders, a coarse aggregate of feldspar, quartz, and some black mica. One of these boulders, which lies on the left side of the creek, must be at least twenty-five feet in diameter, and is partially rounded by atmospheric influences. The other is perhaps six feet in diameter and lies in the middle of the stream, and between the two, in a space of three feet, the greater part of the water of the brook rushes down with considerable force.

The fourth spring is perhaps fifty feet above the second, on the right side of the creek, and within four feet of the water's edge. There is no sediment deposited around it, and, although the water bubbles up somewhat, it is rather chalybeate than otherwise. The taste is scarcely perceptible, and but little notice is taken of it by tourists.

The basin of the second spring is about four feet deep and is used for bathing. The first three springs are strongly impregnated with carbonic acid gas and are the true springs.

These springs must necessarily have their origin in the metamorphic rocks, although the waters may pass up through a considerable thickness of the older sedimentary beds. On both sides of Fountain Creek there is a considerable thickness of the carboniferous beds, but the creek seems to run through a sort of monoclinal rift, though at the falls above, the stream cuts through the ridges nearly at right angles. At any rate, there cannot be a very great thickness of the unchanged rocks below the surface at the springs.

As these springs must at some period become a celebrated and popular resort for invalids from all parts of the world, I will add an analysis of a fragment of the incrustation mentioned above, as given in Fremont's report, page 117.

Carbonate of lime	92.25
Carbonate of magnesia	1.21
Sulphate of lime, chloride of calcium, chloride of magnesia	.23
Silica	1.50
Vegetable matter	.20
Moisture and loss	4.61
	100.00

"At 11 o'clock, when the temperature of the air was 73°, that of the water in this was 60°.5; and that of the upper spring, which issued from the flat rocks more exposed to the sun, was 69°. At sunset, when the temperature of the air was 66°, that of the lower springs was 58°, and that of the upper 61°."—FRÉMONT.

CHAPTER III.

FROM COLORADO CITY TO SPANISH PEAKS.

Looking toward Colorado City from the south, it would seem that the rift, or pass in the mountains through which Fountain Creek *(Fontaine qui bouille)* flows, formed a line of separation between the ranges of mountains; that the north range died out suddenly, in its southern extension, at this point. There is a plain valley of separation visible.

A little below the city, the ridges, or "hog-backs," flex to the southwest and jut up against the base of the mountains and disappear. These mountains are of that abrupt type which I have before referred to; that is, they form the west portion of an anticlinal, the east half of which is not visible. These mountains I call abrupt because the summits are formed of projecting masses of rocks leaning over eastward beyond the base, where this class of mountains occur. The sedimentary beds jut up against the base without any special dip, or, at any rate, there is no wide belt of upheaved ridges, but the most recent formations in the region lap on to the base of the mountains. The immediate eastern range north of Colorado City, and the one south, are, it seems to me, fine illustrations of this statement, and I am more and more convinced that it is correct.

Passing over that portion of the country south of Colorado City, between Fountain Creek and the base of the mountains, the upper cretaceous beds, No. 4, are quite extensive, with *Baculites ovatus* and *Inoceramus* in great quantities. The cretaceous rocks are well shown, especially the upper portions, in the valley of Fountain Creek, from Colorado City to its junction with the Arkansas River. A number of species of fossils, especially shells and saurian remains, are found quite abundantly. There are also scattered about, remnants of the Monument Creek group; and below Colorado City these recent tertiaries occupy considerable area, and reach a good thickness.

But the most conspicuous feature that we observe is the vast quantity of granite boulders scattered over the surface near the base of the mountains, extending at least to Fountain Creek. They diminish in size as they recede from the mountains, and are not much worn.

About ten miles below Colorado City the "hog-backs" appear again faintly in the form of one or two narrow ridges. The lofty mountain, rising up abruptly two thousand or three thousand feet above the base, stops suddenly, and lower granite ridges, with their eastern sides sloping and covered with grass, come in.

About fifteen miles south of Colorado City a little wooded stream that issues from the mountains seems to form the northern limit of a high ridge, which at first extends from the foot of the mountains in the form of a pretty high "hog-back," but soon passes down southeast into the variegated sands of the Monument Creek group. From this point to within a few miles of the Arkansas, the recent tertiary beds are quite prominent. The mountains seem also to be composed largely of igneous rocks.

About fifteen miles south of Colorado City the road to Cañon City passes among the upheaved ridges which form a very narrow belt at first, but continues to increase in width until we come to the valley of the Arkansas, where they spread out to a great breadth.

At the point south of Colorado City where the upheaved ridges reappear, the mountains begin to break up into low hills and fragmentary ranges which continually run out in the plains. Indeed, the entire eastern flank of the mountains, as we pass from the north southward, exhibit an irregular but distinct "en echelon" arrangement; and at a number of localities, the ranges will pass off in the prairies, south or southeast, in groups, thus causing an abrupt notch or bend in the range. There is also in the cañon of the Arkansas an extensive bow or notch, where the upheaved ridges are very conspicuous and numerous, where the complete series of formations, in their regular order of sequence, are thrown up to the vision.

After entering among the upheaved ridges we find the lower cretaceous sandstones forming a conspicuous ridge, inclining thirty degrees to forty degrees about northeast. Then come the variegated sandstones and the brick-red beds inclining at various angles as heretofore described. Before reaching the Arkansas some of the ridges become very large and high, from five hundred to six hundred feet. In very many localities, for a long distance, the red sandstones lie distinctly against the granite hills. Not unfrequently for fifty miles or more along the eastern base of the mountains, all the unchanged beds have been worn away from the metamorphic, and a smooth, grassy valley intervenes, so that it is sometimes difficult to find the two classes of rocks in contact.

About ten miles north of the Arkansas we have an immense ridge, at least eight hundred feet high, capped with lower cretaceous sandstones, and below them fine arenaceous sands, clays, thin beds of limestone passing down into variegated layers, with a heavy bed of gypsum, from fifteen to thirty feet thick, at its base. This bed of gypsum seems to form a sort of dividing line between the brick-red beds and the variegated sandstones above. Passing Beaver Creek we come into a fine oval park, with the large ridge on the east side, and the low red sandstones, which lie on the granite, on the west side. This park is about four miles long and half a mile wide. The bed of gypsum is very conspicuous.

In the vicinity of the Arkansas Valley the cretaceous formations become quite apparent, and while there seems to be no marked line of separation between the divisions, yet portions of Nos. 1, 2, 3, 4 and 5 can be distinctly seen. On Oil Creek, near Cañon City, there are high isolated hills which show the black shales of No. 4, gradually passing up into the rusty arenaceous clays of No. 5. High on the flanks of the mountains can be seen the carboniferous beds, inclining at large angles. The hills are covered with small pines, mostly the piñon, *(Pinus edulis,)* but all the lumber has to be brought from a distance of thirty or forty miles.

High up in the foot-hills of the mountains, in the valley of Oil Creek, a branch of the Arkansas, are the celebrated Oil Springs. There are four of them from which oil is taken, but they are near together, and probably all come from the same source. The oil seeps out through sandstone seventy or eighty feet beneath the surface. A hole has been bored down three hundred feet, but no regular reservoir has been found.

About four thousand gallons of refined oil have been made here per year, for the past three years. There are many impurities in the crude oil: twelve per cent. benzine; fifty per cent. heavy oil; the remainder is tar and nitrogenous matter; much of it is paraffine, and paraffine oil. There is also about fifteen per cent. of useless matter. I saw more than twenty barrels of refuse oil at the spring, which had been rejected from the refinery. This is used for greasing wagons, &c. Specific gravity 38.

The lower cretaceous rocks rise in vertical cut bluffs, four hundred to six hundred feet above the oil springs, and the creek cuts through the upper part of the variegated beds. The course of Oil Creek is nearly south. A range of mountains extends down along the east side of the creek, and runs out before reaching the Arkansas, and on the west side the various formations are shown in a nearly horizontal position, or inclining southwest at a small angle. Indeed, Oil Creek flows through a sort of synclinal valley in part, and near the source of it the red or triassic beds rest upon the granites. All along this creek, where the unchanged rocks are well shown, the lower cretaceous beds seem to pass down into a narrow belt of ashen gray sands and sandstones, which continue down into a variegated series of beds, a part of which I regard as jurassic.

Near the oil springs there are, above the reddish beds, six layers of massive sandstones, varying from ten to twenty feet thick, with seams of arenaceous clays, from a few inches to ten feet in thickness. These rocks exhibit all the indications of shallow water deposition in places, but not a fossil of any kind could be found, and, therefore, it is difficult to determine whether they are lower cretaceous or jurassic.

As to the sources of this oil, I could gain no reliable information. The borings have gone down into the pudding-stones of the lower triassic, and yet no reservoir has been found. It is not known but that the oil may come up from the granites. Great quantities of salt water issue from the springs with the oil, and the oil is taken from the surface of the salt water.

At Cañon City, where the Arkansas comes out of the mountains, on the south side of the river, the principal ridge or "hog-back," which is composed of No. 1, dips 34°, and has a trend about southwest; while on the north side the long ridge, of which there is a very high one, like a lofty wall, composed of the sandstones of No. 1, while a lower outer ridge is made up of the fine calcareous sandstones of No. 2, filled up with *Inoceramus*. It is from this low ridge that the stone for building purposes is obtained. It is not very durable, but works easily and makes handsome structures. This regular wall extends northward, bordering the plain in a straight line for five or six miles, and is very conspicuous.

Issuing from the ground, between the ridges of cretaceous No. 1 and No. 2, in the valley, about a mile above Cañon City, is one of the finest mineral springs we have seen in the West. It is quite small, but the water is delicious. It is doubtless the same, essentially, as the springs at Colorado City.

Just back or inside of this sandstone wall No. 1, is an ashen-gray bed of arenaceous layers, with a bed of fine silicious limestone, containing what seems to me to be indistinct fragments of fresh-water shells. This belt passes down into the red pudding-stones below. Passing up the Arkansas a few hundred yards further, we come to the metamorphic rocks.

About four miles below Cañon City, on the Arkansas River, are some isolated hills, looking in the distance like fortifications, composed of Nos. 4 and 5 cretaceous, capped with a rusty yellow sandstone, which I regard as the lowest bed of the coal formations.

Both the cretaceous and tertiary beds seem to dip southwest five to ten degrees, while on the south side of the Arkansas the tertiary beds incline rather northeast, so that there is an obscure synclinal which shows the influence of the ranges of mountains on each side of the valley. The coal strata have all the characteristics of the older tertiary sandstones, as shown in the Laramie Plains.

Between Cañon City and Hardscrabble Creek, the tertiary beds jut up against the Wet Mountain range, concealing all the older rocks. About half a mile east of Cañon City, the high cretaceous ridges are seen, and then they disappear beneath the tertiary beds, and reappear at the head of Hardscrabble Creek, about thirty miles to the eastward.

High up the foot of the granite hills of Wet Mountain, an obscure synclinal valley can be seen, through which flows a small branch called Oak Creek. The dip of the tertiary beds on either side is nowhere more than ten degrees, seldom more than five degrees. The coal crops out in many places. In the sandstones are the peculiar concretionary forms which are common in these beds everywhere. Their general appearance points out their age to the eye at once.

About ten miles below Cañon City a coal bed has been opened and wrought to some extent. I obtained here the following section of the strata:

9. Sandstone and clay to the summit of the hill –	30 to 40 feet.
8. Carbonaceous and arenaceous clay –	10 feet.
7. Yellowish, gray, soft, fine-grained sandstones –	10 feet.
6. Carbonaceous clay, passing up into laminated clay –	20 feet.
5. Coal – – – – – –	1 foot.
4. Drab carbonaceous clay – – – –	10 feet.
3. Coal – – – – – –	5 feet.
2. Drab clay – – – – –	4 to 8 feet.

1. Yellow ash-colored arenaceous clay, passing down into a yellowish gray sandstone.

In the clay are nodules of iron ore, which are full of impressions of deciduous leaves, like *Salix*, *Platanus*, *Thuya*, and a broad flag-like plant are abundant.

All through the clay there is a yellow powder, oxide of iron, and seams of gypsum. Much selenite is scattered through the beds of clay and coal. The plants, so far as I have seen, are found in the clays just above the coal.

The yellow arenaceous clays of No. 5, in the Arkansas Valley, pass up into a somewhat extensive series of what I call mud beds, composed of alternate thin layers of clay and mud sandstones, with all kinds of mud markings, sort of transition beds or beds of passage. In the upper portion of these layers I found an imperfect specimen of *Inoceramus.* This group of beds is from fifty to one hundred feet in thickness. Resting upon them is a thick bed of rusty yellow sandstone, which I regard as the lower bed of the tertiary deposits, and marks their commencement in the Laramie Plains, on the Arkansas River, and the Raton Mountains. Below these beds of passage there is a yellow, arenaceous, marly clay, full of iron-rust concretions, with an abundance of small bivalves and other shells, with *Baculites ovatus*—plainly No. 5.

It is now clear that the Cañon City coal formation occupies a very restricted area; that the entire thickness of the beds cannot be more than six hundred to eight hundred feet; and that it is an isolated portion, protected from erosion in a manner not easily explained, and that it was once connected with the same formations in the Laramie Plains, about Denver; southward in the Raton Mountains, and most probably also with those containing coal in the valley of the Rio Grande. The area occupied by the coal beds lies east of Cañon City, between Wet Mountain and the Arkansas River, with the eastern limit three or four miles before reaching Hardscrabble Creek. It is about twenty miles from east to west, and five to eight miles wide from north to south; and only a small portion of it will furnish coal. The coal itself is quite good for the purposes of fuel, but the beds are not thick, and the quantity is not great. There is the usual quantity of brown iron ore connected with these beds.

The Arkansas River flows through the synclinal depression, below the mouth of Hardscrabble Creek.

It may be that the older rocks are elevated under the debris close to the foot of Wet Mountain, but no beds older than the cretaceous can be seen. The upper cretaceous beds extend up close to the mountains, oftentimes capped with the tertiary, inclining not more than five to ten degrees.

At the head of Hardscrabble Creek the ridges of upheaval or "hog-backs" begin to show themselves again in a narrow belt which rapidly widens out, so that before reaching Greenhorn Creek they have spread out to a width of several miles.

On Red Creek, which is about eight miles south of Hardscrabble, there is the finest exhibition of the yellow massive chalk passing down into the gray marl of No. 3 that I have seen south of the Upper Missouri. In the channel of this stream and its branches there are vertical walls eighty to one hundred feet high, looking much like irregular mason work. Some of the gray portion is a very hard limestone, and contains a large, apparently undescribed species of *Inoceramus*. Between Red Creek and St. Charles there are other exhibitions of the cretaceous rocks, but especially of the quartzose sandstones of No. 1, which, at the foot of the mountains, are cut through by the numerous little branches in a most picturesque manner. The little streams run through vertical walls eighty to one hundred feet high, forming most interesting cañons, and revealing all the peculiarities of structure of this sandstone. Some of it is coarse and friable, other portions are compact silicious rocks; others, a pebbly conglomerate. All the illustrations of irregular layers of deposition, ripple or wave markings peculiar to sandstones, are found here; also, admirable examples of slickensides. The jointage, which is very marked, is vertical, at right angles to the lines of stratification, and most essentially assists atmospheric agencies in wearing it away, so that the sides of the walls are often very rugged, and immense cubical blocks have fallen down by the water's side.

The different formations all along the flanks of the mountains are exposed by the upheaval of the mountains, and lie in belts or zones, which are sometimes concealed for a distance by recent tertiary deposits or by debris; or they are narrow or wide at different points, and their conditions are only to be determined by personal inspection.

At the head of St. Charles Creek all the rocks incline gradually down from the mountain side. No. 1 dips thirty degrees and slopes gently down until it reaches a nearly horizontal position in the plain. West of this first high ridge is a fine valley in which are beautiful, cultivated farms. The red beds are well shown, and I have no doubt but that the carboniferous limestones are exposed on the sides of the mountains.

Just before reaching Greenhorn Creek all the small ridges and the first high one run out in the plain, and the mountains flex around toward the southwest to form the notch for the Sangre de Christo Pass. The ridges of elevation and the side ranges, like Wet Mountain, have a general trend about northwest and southeast, and all the lower ridges run out in the prairie, and Wet Mountain ceases at the pass.

On the north side of Greenhorn Creek, near Hicklin's Ranche, No. 2, is a rusty arenaceous limestone, full of shark's teeth, mingled with a small species of *Ostrea*. The arenaceous limestone is attached to a gray, fine-grained sandstone, and is rather concretionary in form. Just on the opposite side of the creek, and apparently holding a higher position, are the quartzose rocks of No. 1. Around the south end of the Wet Mountain, the cretaceous beds, Nos. 2 and 3, in the form of dark clay, and yellow, chalky shales, present bench-like hills, extending down at right angles to the strike of the range or eastward, and present an east front with nearly horizontal strata.

All the ridges along the flanks of Wet Mountain have a general strike of northwest and southeast, and run out in the plain. Wet Mountain also flexes around slightly so as to end nearly or quite in a south trend, while the ridges appear again on the southwest and west side, running up into Huerfano Park. Here we see on the west side of the Wet Mountain range, the red beds and cretaceous formations, corresponding to those on the east side. The park is largely occupied with the calcareous shales of No. 3.

Just before reaching Badito, in the Sangre de Christo Pass, there is a long ridge, extending down westward from the Wet Mountains, which is composed mostly of the red and white sandstones of the triassic, inclining twenty-five to thirty degrees. At Badito we find mostly a reddish-gray quartzose sandstone like No. 1, and it forms the foot-hills of the mountains. As usual the dip of the bed is in various directions and at different angles. The Huerfano Creek is a fine stream with a moderately wide valley which is all cultivated by Mexicans. Huerfano Park is about fifteen miles long and from three to five wide, and is already filled with settlers. It is surrounded on all sides by mountains composed of igneous and metamorphic rocks. Black Butte, the principal peak of Wet Mountain range, appears perfectly round or mammi-form and is basaltic. Scattered over the area of the park are several outbursts of basalt. The cretaceous beds dip south in some places ten to twenty-five degrees; in others they are nearly horizontal. As we ascend the pass by the road we can see three considerable ranges called the Veta Mountains—one range on the north side and two on the south side—all igneous rocks. They all have sharp sierra-like summits.

These dikes have so heated the sedimentary rocks in their vicinity that we have here every variety and grade between unchanged and changed rocks. The summits and sides of these mountains are covered with a continuous mass of debris of broken rocks, and this mass has the appearance of being just ready to fall down, like an immense land-slide. On the sides of the mountains near the pass are belts of quartzose sandstone, some of it a pudding-stone—really forming a portion of those seen on the west side, for I do not think we come to the axis here until we find the granitic belt, some eight or ten miles west of the immediate summit of the pass. We therefore have the cretaceous rocks, limestones, and sandstones, and then the reddish sandstones at the summit, and then farther west the full series of carboniferous limestones. From the divide between the Greenhorn and Cuchara creeks, looking southward, is one of the most extended and beautiful views on our route. The long level benches extend down from the mountains, apparently breaking off from point to point, and appearing high at the place broken. These benches are planed off so as to look like long tables, and, with the valleys between them, seem to me to show clearly the direction of the eroding force. All these benches are underlaid by the soft sandy marls of Nos. 2 and 3, cretaceous.

Huerfano Butte rises up in the midst of the plain in the valley of Huerfano Creek. The rocks are basaltic, some portions a true syenite. It is evident that it is a portion of a dike which has extended northeast from the mountains. Much of the rock is massive igneous granite. Fragments of cretaceous clays, changed by heat, are scattered around the butte. It seems to me that this is a dike, thrust up before the superincumbent beds were swept away, and that the igneous material never reached the surface in a melted state. The butte is about two hundred feet high, the rocks being of a dark steel-gray color. There is no evidence that the underlying strata have been disturbed by this butte.

The evidences of igneous protrusions are everywhere abundant, south of this point, for two hundred miles. The Spanish Peaks I regard as a gigantic dike, with the strike about northeast and southwest. The entire surface of the country, from the Spanish Peaks to the Raton Mountains, is penetrated with dikes, which often reach far across the country with a trend about northeast and southwest. The cretaceous rocks are in many places much changed by contact with the fluid mass, and in some cases the strata are somewhat disturbed. The clays are turned into slates and the sandstones into dark steel-colored rocks, much like the basalt itself. In No. 2 I found a species of *Inoceramus*, very distinct, and a *Modiola*.

About ten miles before reaching the Apishpa Creek the tertiary sandstones begin to show their abrupt bluffs on our right. I am convinced that beds of this age entirely surround the Spanish Peaks and the mountains in the vicinity. This abrupt front continues north of the Raton Mountains until we come to Trinidad, and presents a singular feature in the scenery. It would seem to form a sort of a shore line of a wonderful basin, as if a body of water had swept along and washed against these high bluffs, as along some large river. That these beds once extended far out into the plains eastward, seems clear, and the evidences of erosive action are enormous. Here, abrupt bluffs which form these different shore lines are four hundred to six hundred feet high above the creeks, and the dip of the strata is about five degrees west or southwest. In the plains to the eastward are isolated mesas, which are left as monuments to show that these beds, with the igneous outpourings, once extended over a large part or all of the space to the eastward, which now looks so finely leveled off like a meadow. This wall-like front extends sixty or eighty miles in a nearly direct line southward, capped with a thick bed of basalt, for the most part.

Just east of the Spanish Peaks a distinct synclinal can be seen in the tertiary beds. They dip slightly from the peaks, and from the bluffs they dip gently toward the peaks, enough to produce a distinct depression of considerable length. I do not know why the tertiary strata incline toward the mountains, unless they have been partially elevated by the dikes.

As far to the southward as the eye can reach, the country looks rugged and mountainous, with some curious mesa-like summits covered thickly with the *piñon*. These tertiary beds are composed as usual of alternate beds of rather yielding sandstones of all textures and composition, with clays, some of which are carbonaceous. The harder beds project out from the sides of the hills, while the softer beds are smoothed off and covered with grass or other vegetation.

Near a stage station, about ten miles south of Apishpa Creek, the cretaceous clays, No. 2, are cut through by a small creek, so as to reveal three dikes within the space of thirty feet. The first is well defined; four inches wide, vertical, looking like a stratum of dark brown sandstone standing perpendicular; strike twenty degrees north of east. Second dike, strike northeast and southwest; four feet wide. Third dike, northeast and southwest, from twelve to eighteen inches wide. The clays are not disturbed, and are perfectly horizontal, but so changed on each side of the dike that the cleavage has the appearance of stratification. I am convinced that in the case of these small dikes the melted material has been thrust up through the cleavage openings. There are very many dikes in this region, all of which have a similar direction. I suspect that in all cases of dikes these cleavage openings are lines of least resistance, and form the apertures for the exit of melted material, and that the surrounding strata are not disturbed only where the pressure from beneath is too great. I would simply suggest, however, that it is quite probable that as there are in nearly all rocks two sets of cleavage lines crossing each other at certain angles, so there are two sets of gigantic cleavage lines for the earth's surface, which have formed the lines of least resistance to the elevation of the mountain ranges—the basaltic ranges in most instances having a strike northeast and southwest, while the metamorphic ranges trend northwest and southeast. The eruption of the igneous rocks is an event subsequent to the elevation of the metamorphic ranges. Sometimes the eruptive rocks seem to trend northwest and southeast, or nearly so.

On the hills surrounding Trinidad are great quantities of impressions of deciduous leaves in the rocks. The most conspicuous, as well as abundant, fossil, is a species of fan palm, undoubtedly *Sabal campbellii*, which occurs in the lignite beds on the Upper Missouri. This plant would seem to have formed the dominant tree in ancient times, much like the palmetto of South Carolina. In some places the calcareous sandstones are filled with this plant for miles. There are also, in considerable abundance, leaves of the *Magnolia*, *Platanus*, *Laurus*, &c., and, so far as I can determine, identical with the species found on the Upper Missouri. I do not doubt for a moment that all the coal beds of the Raton Mountains are tertiary and belong to the great coal system which has already been traced over such a wide area. In a little dry creek I observed an outcrop of coal, about two feet thick, with drab clay above, filled with brown iron ore, and above this a gray laminated sandstone. In this sandstone a huge specimen of the *Sabal* was found.

About four or five miles up the Purgatory River, above Trinidad, on the south side of the creek, I examined two openings that have been made for coal. It is the same bed in both places, and is about four or five feet thick at the outcrop. Underneath it, is a sort of indurated sandstone with very irregular laminæ, with thin layers of vegetable matter. Immediately beneath the coal is four to six feet of drab arenaceous clay, with large concretionary masses of iron ore of excellent quality; above the coal is drab clay passing up into sandstone. These openings for coal are about fifteen feet above the bed of the creek, and the strata are nearly horizontal. The clay above the coal at the other opening, not far away, is, perhaps, eight feet thick, and full of iron ore, with leaves like willow and nuts and small filiform leaves like grass. The clay is a drab steel color passing gradually up into a very rugged sandstone with projecting hard layers, which give a wall-like appearance to the bluff-like sides. From the Spanish Peaks to Trinidad, and along the Purgatory Creek for four miles above, the black shales of the cretaceous are visible. Usually in this region these drab shales pass into a series of alternate clays and sandstones in thin layers, and upon them rests a conspicuuso bed of rusty yellow sandstone, which I have regarded as the lowest bed of the tertiary series. A bed of sandstone precisely similar to this, and holding the same geological position, occurs at Cañon City and the Laramie Plains. But at these localities the intermediate cretaceous beds, Nos. 3, 4, and 5, are not absent, while in the Raton Mountains the sandstone seems to rest directly upon the lower cretaceous formation, No. 2. I have searched this sandstone over an area of many miles for fossils, and I only succeeded in finding one obscure fragment of a marine bivalve like the clam, while in the mud beds and shales below, specimens of *Inoceramus* are common. I make this sandstone, therefore, the line of separation between the tertiary and cretaceous formations in this region. If this is true—and I am confident that it is—there is an entire want of continuity in the sequence of the beds.

In a dry gulch, about two miles west of Trinidad, there is a bluff with about thirty feet of black cretaceous shales, No. 2, with an irregular surface, on which is deposited ten to fifteen feet of partially worn pebbles, held together by a carbonate or silicate of lime. Much of it looks like tufa. In this place there is quite a deposit of what appears to be the excrement of birds or bats, but which has been oftentimes mistaken for the indications of petroleum. This deposit of pudding-stone seems to be quite common, and is well shown in the banks of all the dry creeks.

Raton Peak is the highest point in this region, and I have estimated it to be about eight hundred to a thousand feet above Purgatory Creek. It is capped with a huge mass of basalt, and underneath it is a great thickness of the tertiary strata, some layers of which are full of impressions of leaves. I distinctly recognized *Sabal*, *Platanus*, *Carya*, *Cornus*, and *Populus*. In the muddy sandstones, just underneath the coal bed, are an abundance of a species of pine cone in the form of casts.

Crossing the road, about four miles west of Trinidad, is a beautiful illustration of a dike, about twelve to fourteen inches wide, with a strike twenty degrees south of east, and a slight inclination southward. It is thrust up through a considerable thickness of the lower tertiary beds. The rock seems to be very heavy, though full of cavities, filled with a whitish substance which cuts easily with a knife—calcite or carbonate of lime. The hills north of Fischer's Peak, through a bed of coal. A little further the mass of the rocks has a rather bright, black color. This dike runs along the road and passes over another dike, which is more obscure and not as well defined. On the east side of the road are several outcroppings of coal in the sides of the hills. The coal is about four feet thick, with arenaceous clay above, passing up into sandstone.

About five miles south of Trinidad, on the east side of the road, is another exposure of the coal in the banks of a little creek, which is worthy of notice. From the water's edge up there are layers of fine-grained sandstone filled with bits of vegetable matter. Above this comes a bed of black shale, four feet, passing up into a gray sandstone, rather concretionary and irregular in its line of deposition. This bed is fifteen or twenty feet, sometimes solid sandstone. Then in a little distance it will be separated by a bed of shale or black slate. Above the sandstone is shale with iron ore; then about two feet of mud sandstone; then very black clay, nodular in some places—the middle portion impure, earthy coal—five feet; then two feet laminated bluish gray sandstone, with stems and bits of vegetable matter scattered through it; then black coaly shale, —eighteen inches; passing up into a layer of good coal—twelve inches; black shale—four feet; then a layer of sandstone—three inches; then black shale passing up into arenaceous clays; then black shale—six feet; then a bed of coal—six or seven feet. Immediately above the coal bed, without any clay, is an irregular gray, rusty sandstone, full of concretionary layers, and readily yielding to atmospheric influences. Then comes drab arenaceous clay—three feet; good coal—four feet; drab arenaceous clay, with very large concretionary masses of brown iron ore. This clay bed must be fifteen or twenty feet thick, passing up into a soft yellow sandstone, fifteen or twenty feet thick, and capping the first hill. Then alternations of sandstone and clays continue far up the distant hills for hundreds of feet, until we reach the mesa or basaltic cap. Here some coal beds show plainly along the road for six or eight miles above Trinidad, and still higher up on the hills, now concealed by vegetation, I have no doubt that there are beds of coal. The mesa, of which Raton and Fischer's Peaks form parts, is undoubtedly the overflow of a dike, which seemed to take a general direction northeast and southwest, and toward the northeast appears to incline about ten degrees.

In ascending the Raton Mountains by the road the cretaceous beds soon disappear; the tertiary come in with coal and soon disappear in turn. The dip of these beds I found difficult to determine, and, I think, when there is any, it is local, and that in the aggregate they may be regarded as nearly horizontal. Just before reaching the toll-gate, near Mr. Wooten's, the sandstone inclines northward about fifteen degrees. Near the toll-gate, by the side of the road, a bed of impure coal, two feet thick, has been exposed. In a ravine further south there is an opening from which coal is taken for fuel, the bed being four feet thick and of excellent quality. This bed has some impure coal above and below, and when opened I think that it will prove to be from six to eight feet thick, good coal. The grass and debris so cover these hills that it is impossible to get a connected section of the beds, but the usual clays and sandstones occur above the coal.

Toward the southern end of the pass there are some perpendicular walls of sandstone which show a vertical cleavage, strike southeast and northwest. In this sandstone are two or three small seams of coal, two to four inches thick, which break the lines of cleavage and interrupt them. This sandstone is from one hundred to one hundred and fifty feet thick, and immediately beneath it is an irregular bed of the alternate thin layers of the mud sandstone and clay, which I have called a bed of passage between the cretaceous and tertiary of this region. I call it a sort of mud shale, as the sediments seem to indicate a continuous mud flat, with the surface of the sandstones and shales covered with all sorts of mud markings. As we emerge from the Raton Mountains southward to the plains we find a large thickness of this mud shale with the sandstones above. There seems to be three hundred to four hundred feet of sandstone, with a cap of basalt. At the foot of the hills there is a dike with a strike northeast and southwest, with a width of about six feet. This dike is shown on the west side of the road in the form of a pile of horizontal columns, like cordwood, fifty feet high or more. Some of the columns are five-sided, but mostly four-sided.

All along our right hand the high hills are precisely as they were from Spanish Peaks to Trinidad. These bluff-hills continue like an irregular wall as far as Maxwell's. They are cut up by side streams into cones and ridges, giving a wonderful picturesqueness to the scenery. This range of hills presents the same kind of shore-line as is seen north of the Raton Hills, with the lower cretaceous shales and the sandstone in juxtaposition. On the east side of the road, broken portions of these ridges extend down southward or southeast. Scattered over this broad plain are buttes and mesas—isolated exhibitions of the basaltic rocks. The tertiary beds soon cease in the plains to the eastward, and the cretaceous beds occupy the country. That all this beautiful valley or plain on the east side of the Raton Hills has been carved out of the tertiary strata appears to me most probable. Why the eroding agency left such a belt of hills as the Raton it is difficult for me to determine, but I am disposed to believe that it acted from the northwest toward the southeast, and was local. The direction of all the benches of cretaceous material left in the valley, as well as that of the mesa tops, has this general trend, and the map will show the numerous branches which flow from the mountains into the Canadian River through these tertiary hills. I have called the bluff-hills on the west side of the road a shore line, because they present almost vertical sides like the bluffs along the Missouri River. These hills show first lower cretaceous shales, from one hundred to one hundred and fifty feet thick, then fifty to one hundred feet of sandstone, the coal beds overlaid with sandstones again. When any of the little streams cut these beds, they reveal the coal, as in the Vermejo Creek and others, ascending toward their sources.

Near the Vermejo Creek I obtained the following general section ascending:

1st. Cretaceous shales, with *Inoceramus* and *Ostrea*.

2d. Massive heavy bedded sandstone, yellowish gray, rather concretionary in its structure, and weathering by exfoliation.

3d. Three thin seams of coal, with clay above and below, in all twenty feet in thickness.

4th. Rusty gray sandstones, fifteen feet.

5th. Clay, passing up into a thick bed of coal, apparently from six to ten feet thick.

6th. The coal is overlaid immediately by a soft sandstone, which passes up into a heavy bedded sandstone, fifty to eighty feet thick.

7th. One hundred and fifty feet of arenaceous clays, two beds of coal about midway, one twelve inches the other four feet thick, with a few thin beds of sandstone.

8th. Capping the hills is a bed of sandstone of indefinite thickness.

In the sandstone are immense rounded masses of a deep, dull reddish, rather fine-grained sandstone, which is evidently concretionary. Many of these masses have fallen down on the sides of the hill, and are now disintegrating by the process of exfoliation. From these high hills one can look with a field-glass fifty to one hundred miles into the plains southeast, along the valley of the Canadian River. A long, mesa-like ridge extends down from the mountains and finally dies out in the plains. I am confident that the conical hills on the north side of the Vermejo are six hundred feet above the bed of the creek.

I am now satisfied that these tertiary strata extend close up to the mountains from the Spanish Peaks to Maxwell's, and the only way I can account for the very slight disturbances of the sedimentary beds is the fact that the mountains to the west of them are mostly basaltic. The miners in the Moreno Valley regard it as very strange that gold mines and coal beds should be found in the immediate vicinity of each other. From the fact that these hills or mountains are composed almost entirely of horizontal strata of comparatively recent date, I think they should be called simply hills. They occupy quite an extensive area, and contain a vast quantity of coal and iron ore, practically inexhaustible, however great the demand in future years. The brown iron ore of this vicinity is the richest I have ever seen in the West, and the coal is equal to any ever discovered west of the Missouri river, except that in the Placiere Mountains of New Mexico. Between the Cimarron and Rayada Creeks, a lofty ridge, one thousand feet or more in height, extends from the mountains with a trend a little south of east, the dip north about forty-five degrees. North from this ridge, which is composed of altered sandstones, the tertiary beds dip gently about five to ten degrees. Between these and the altered sandstone ridge is a cretaceous ridge, five hundred feet high, inclining at a moderate angle. This ridge of altered sandstone seems to be a sort of side elevation or spur, prolonged eastward from the main range, and soon ceases.

From Maxwell's to Fort Union the plain country is occupied by cretaceous rocks, mostly the dark shales of No. 2, though the sandstones of No. 1 appear now and then, especially in the vicinity of Fort Union. Scattered all over this broad space are a vast number of conical buttes and mesas, so that the surface would seem to be pierced everywhere by dikes or outbursts of basaltic rocks. Since leaving the Arkansas River southward, the cretaceous formations seem to have increased greatly in importance, and here No. 2 seems to be enormously developed. After leaving the Cimarron southward, a peculiar configuration of the surface commences, which has been gradually unfolding ever since we left the Spanish Peaks. From this point to the Cimarron there was a commingling of features, those that result from the outpouring of the igneous rocks, and those from the weathering of the tertiary strata. South of the Cimarron, the vallies are more narrow and more sharply defined, as are the cones and mesas, and the only formations involved, so far as the plains are concerned, are the igneous rocks and the lower cretaceous. The grass is excellent in the vallies, and the hills are covered with *piñon*. No good timber is found anywhere, so that the *adobe* method of building houses adopted by the Mexicans would seem to have arisen from the natural deficiencies of the country. The mingling of the eroded material of the igneous rocks with the cretaceous clays, sands, and marls, seems to have produced a good soil. The vallies appear to have been carved out of the basaltic mesas, sometimes with wonderful regularity and beauty. There are several sets or series of mesas, as it were. The higher mesas are covered with a great thickness of basalt with vertical sides, the basalt breaking into columnar masses. The lower mesas seem to be more level or table-like, and are covered thickly with fragments of basalt. It is quite possible that these different mesas represent different levels of the surface, prior to the outpouring of the fluid material. Between Sweetwater Creek and Ocaté, I found near the road some yellow sandstones, filled with fragments of *Ostrea*, which I think belong to the upper part of No. 2.

Near Ocaté, the peculiar carving out of the valleys by erosion is seen, presenting to the eye the most beautiful views that can be conceived of in the natural world. They seem to have been formed by the hand of art. No other condition of the surface could have admitted of their existence. The fluid material seems to have been poured out over the surface in one continuous and almost uniform sheet or layer, and these valleys are thus carved out of the mesas. The little streams cut narrow channels through these basaltic plains, sometimes very deep and often for miles without a bush to mark the water course, so that they are not observed by the traveler until he is in close proximity to them.

From Ocaté Creek to Fort Union, the surface is covered with volcanic rocks, many of which are so porous as to seem like pumice. These masses are so light that they must have been scattered by the wind. There are great numbers of hills and ridges scattered in every direction, covered thickly with these igneous fragments.

CHAPTER IV.

FROM FORT UNION TO MORA.

Through the kindness of Dr. Peters, United States Army, the surgeon of Fort Union, I made a short journey to Mora Valley, about eighteen miles west of Fort Union, and I am also indebted very much to him for his knowledge of this country.

About due west of the fort is a long ridge which runs nearly north and south, and is underlaid by the quartzose sandstones of No. 1. This ridge is cut through in every direction by dry creeks, which show that the strata are quite horizontal. The plateau or mesa-like summit is about five miles across, when we descend into a park-like area eroded out of the brick-red beds in the Cayoté Valley. The rocks of the plateau are here seen to incline east from five to ten degrees, just revealing the upper portion of the brick-red beds. This valley is about three miles wide and perhaps five to ten miles in length, and at the south end the creek cuts through the cretaceous plateau, forming a narrow gorge. On the west side we have the red upheaved ridge well shown, and all through the valley are fragments of low ridges inclining at moderate angles. Between the little branches of the creek and all around the borders of the valley are well defined terraces. This valley or park is beautifully grassed over, and the benches or terraces are as smoothly rounded off as they well can be. The surface is covered with water-worn bowlders and drift. On the west side of this valley the road passes through the gorge of the Mora Creek, and for nearly ten miles we travel across the upturned edges of the sedimentary rocks. There seem to be here two well defined series of red sandstones; the upper series we have described as underlying the park-like valley of the Cayoté Creek, about three miles wide, and separated by lofty ridges of yellowish, gray sandstone on the east side; and then, west of the gorge, a second series of rather dull purplish or dull brick-red sandstones, all inclining in the same direction but at different angles. The low ridges of the upper series of red beds incline west fifteen, twenty, and thirty degrees. The highest ridge is composed of the yellowish gray sandstone that separates the two series of red beds, and is about one hundred and fifty feet high, and inclines thirty-three degrees.

Passing up the valley of the Mora the sandstones are of all colors and textures, some of the ridges very fine, compact; others coarse-grained, and yielding readily to atmospheric influences; others composed of an aggregate of particles of quartz and small water-worn pebbles. Among the pebbly sandstones there is a thin layer, perhaps a foot thick, of an ashen-gray brittle limestone. This second or lower series of reddish sandstones extends nearly two miles, dipping fifty to sixty degrees; in a few cases nearly vertical. The intervals between these ridges, which are usually from ten to one hundred yards wide, are grassed over and sometimes reveal the fact that they are underlaid by soft shale. Neither in the first or second series of red beds was I able to detect any organic remains.

Within about three miles of Mora Valley we come to a series of alternate ridges of sandstones, limestones, and shales, inclining forty to fifty degrees. The first bed of limestone is full of fossil shells, *Productus*, several species, *Spirifera subtilita*, *S. triplicata*, &c. Then comes a bed of micaceous sandstone, full of vegetable impressions of the genus *Calamites*, and large fruits or nuts. These beds incline sixty-five degrees. After this comes a coarse reddish sandstone, an aggregate of particles of quartz and worn pebbles, most of it a fine pudding-stone. Then comes about three hundred feet of reddish sandstone, then cherty limestone, with *Productus*, *Spirifera*, and other species of true carboniferous types. Alternate beds of sandstone, limestone, and shale continue nearly to the Mora Valley—the beds of sandstone forming about nine-tenths of the thickness. From Fort Union to Mora, eighteen miles, we pass directly west, at right angles, to the mountain ranges, and over the upturned edges of the sedimentary beds from the lower cretaceous to the metamorphic rocks. The sedimentary rocks all incline in the same direction, at various angles, from five to seven degrees. I cannot see that in this vast series of ridges, any beds have been repeated, and, therefore, there must be exposed here in a curiously consecutive manner from eight thousand to ten thousand feet, at least, of sedimentary rocks. The junction of the unchanged rocks with the gneissic beds is rather obscure, but a bed of limestone seems to incline against them. From my observations from Las Vegas to Santa Fé, I am satisfied that all along the mountains the carboniferous limestones rest directly on the granitic rocks. The valley of the Mora, in which the town of that name is situated, is one of the most fertile and beautiful that I have ever seen in the West. It is almost entirely surrounded by mountain ranges, and in the aggregate it forms a high quaquaversal—that is, the rocks seem to incline from all directions toward a common central point. It is about ten miles long from east to west, and two miles wide from north to south. It is in the form of a cross. At the east side is a long valley extending five miles or more in each direction north and south from it. Either one of these valleys, taken separately, would form a synclinal. The whole valley is in part worn out of the gneissic rocks. Mora Creek runs directly through it and every acre of it is under cultivation, and with the rude Mexican style of farming, produces most abundant crops.

All around this valley the slopes of the mountains show clearly that the metamorphic rocks incline from it at very high angles; and all around the borders are foot-hills or low ridges, the remnants that are left after erosion, which show distinctly the direction of the dip. But the series of gneissic beds on the east side of the valley are very interesting, consisting of alternate beds of black banded gneiss, and a coarse aggregate of feldspar and quartz. Some of the beds are composed of mottled gneiss. These beds all incline to the west or northwest, at various angles from twenty degrees to thirty degrees. This series of gneissic strata extends nearly half a mile, and is plainly a remnant left after erosion. They incline in an opposite direction to the unchanged rocks—that is, there is no conformity. This is one of the most interesting points on our route in a geological point of view, and I regretted very much that I could not remain a longer time.

About northeast from Fort Union there is a small range of mountains of some interest, called Turkey Hills. They seem to form a regular upheaval with a line of fracture nearly northwest and southeast, and apparently independent of the volcanic forces that have once operated all around it. This mountain is well covered with timber, and the highest points rise fifteen hundred to two thousand feet above the level of the plain at Fort Union. Entering the mountains nearly north of the fort, we pass up a sort of anticlinal valley; the beds inclining in each direction at a small angle. None but the lower cretaceous sandstones and a portion of the upper series of red-beds are exposed anywhere in this range, which is about twenty-five miles long and ten miles wide. Among the red-beds are two or three layers of bluish limestone, and underneath the cretaceous is a bed of fine-grained whitish sandstone, which I am inclined to regard as jurassic. From the summits of these mountains we can see the Spanish Peaks, Raton Mountains, and, indeed, the whole country round about for a radius of one hundred to one hundred and fifty miles. About nine miles east of Fort Union there is an old volcanic crater of great interest. This is the nearest approach to recent volcanic indications that I have ever seen, or known on the east side of the mountains. The rim of the crater is circular and well defined, though the depression is very shallow. Yet, as we ascended the high volcanic mountain, we found the sides covered with masses of rough basalt, so much so as to render traveling difficult and very laborious. But inside of this crater there is scarcely a rock to be seen, and the slightly concave surface is thickly grassed over. The immediate sides of this mountain all around are covered with longitudinal ridges of the rock which was evidently poured out of the crater and ran down the sides. The circular crater is about fifty yards in diameter, and is now filled up with earth. This rounded mountain must have been built up by the continued overflow of melted rock, and at this time its summit is at least twelve hundred to fifteen hundred feet above Fort Union. In the vicinity are what seemed to be rifts, which have now formed valleys or gulches, and on each side of which are thick borders or walls of the basalt.

About fifteen miles north of Fort Union there is another of these craters which has attracted attention. The depression is about two hundred and fifty feet deep and five hundred yards in diameter, and the rim is broken away on the north side. The borders of the crater are elevated about twelve hundred feet above the fort. This vast mountain mass must be the accumulation of the outpoured melted rocks. All over the sides are immense ridges or banks, as it were, of the melted rock which has flowed out of the crater. The summit is covered with lava, some of it black and some of it of a red color, but very porous and light, like pumice, so that the wind has distributed great quantities for a long distance over the plains below. This melted material has been poured out over the cretaceous beds, often concealing them over large areas. I am convinced that at one period a very large portion of this country was covered with these craters, but none of them seem now to be so well defined as those described.

About four miles north of the fort is a mesa capped with basalt, which is underlaid by cretaceous rocks. Sometimes the basalt is worn away over large areas, uncovering the rocks below. The mesa is about three hundred feet above the fort. The valley in which Fort Union is located is a very beautiful one, and is plainly carved out of the cretaceous plateau. On the west side the abrupt walls can be seen for miles, but on the east the ascent up to the foot of the tertiary mountains is gradual, though here and there the cretaceous rocks crop out.

Before closing this chapter I wish to offer my most cordial thanks to the officers of Fort Union, for courtesies and aid which enabled me to perform the work of a month in a few days. Under the intelligent guidance of Captain W. R. Shoemaker, I spent two most profitable days examining the country in the vicinity of Fort Union, and with Dr. D. C. Peters, United States Army, visited the beautiful Mora Valley. The entire party were the recipients of favors at this post, which showed more clearly than I can express it in words the deep interest which the officers of the army everywhere take in the development of the material interests of that portion of the West where they are stationed. We could also measure the amount of life in the citizens of any town we visited, by the interest they took in our efforts to study the resources of the country. Mr. C. W. Kitchen, especially, and the citizens of Las Vegas generally, extended every attention to us in their power, and I am convinced that at no distant day this must be the most pleasant and prosperous town in New Mexico.

CHAPTER V.

FROM FORT UNION TO SANTA FÉ.

We left the hospitable post of Fort Union with regret and pursued our way southward towards Las Vegas. The first eight miles we passed over quartzose sandstones of No. 1, and then appeared above them a hard bluish limestone, which belongs to No. 2. The sandstones o : No. 1 gradually disappear, and the limestones take their place. Several species of *Inoceramus* occur, and Mrs. General Grier has in her possession an *Ammonites* that came from this region, which is tuberculated like *A. percarinatus*. All the way to Las Vegas we have a fine view of the country along the base of the mountains. The exposures of the sedimentary rocks are wonderful in their extent along the eastern base of the mountains, from Fort Union to the point below Santa Fé, where the range passes out and is lost in the plains. The belt of upheaved ridges is from four to eight miles wide. All around Las Vegas, in the plains, the blue limestones, passing up into an enormous thickness of the black shales of No. 2, is everywhere seen. The little streams cut deep channels through it.

The finest section of the sedimentary rocks of this region, that I have ever seen, may be found between Las Vegas and the Hot Springs, on Gallinas Creek. The beds from the metamorphic to the cretaceous, inclusive, are so regularly and clearly exposed along this creek that it is not possible to mistake their continuity, and I would call the attention of all travelers visiting this country, who have any interest in the geology, to it.

The Hot Springs, which have already become so celebrated for their supposed curative qualities in certain diseases, are located about five miles northwest of Las Vegas, just at the junction of the carboniferous and the gneissic rocks. The lowest spring issues from the granite just underneath a mass of limestone. The bed of limestone that rests directly on the granites is quite hard and cherty, with a dip nearly southeast 40° to 45°. The metamorphic rocks below are rotten gneiss. From this point outward towards the plains I made the following section, passing over the upturned edges as they were exposed with wonderful clearness and consecutiveness to the eye:

1. Hard grayish cherty limestone, resting, directly on the gneiss.
2. Micaceous sandstone full of iron, partly a very micaceous rotten shale.
3. Yellow limestone with less chert, excellent for lime, containing *Productus*, two or three species, *Spirifera subtilita*. Between the beds of limestone, that vary from four to twenty feet thick, are two beds of rusty clay, each four to six feet thick, the whole dipping 50°.
4. Black shale with thin layers of a sort of arenaceous mud, from one-quarter of an inch to four inches in thickness.
5. Limestones with *Productus*, *Spirifera*, corals and crinoidal stems, passing up into a very cherty limestone, one hundred and fifty feet thick; dip 60° to 75°. Among the layers of limestone are thin seams of shale.
6. Grayish brown arenaceous limestone passing up into a somewhat micaceous sandstone—30 feet.
7. Variegated greenish, reddish, ashen, and yellowish shaly clays—20 feet.
8. Variegated sands and sandstones of all degrees of fineness. The prevailing color red, varying from bright brick-red to purple, with some

whitish, yellowish, &c.; dip 45° to 55°; thickness two hundred and thirty-five feet.

9. Rather fine grained grayish sandstone. This bed has passed a vertical position so that the dip is southwest 75°; thickness fifty feet.

10. Variegated sands, light brick-red, dull purple, reddish brown and light gray. The dull purplish sands, ten feet thick, are amygdaloidal, full of almond-shaped nodules and cavities.

10. Alternate beds of light yellowish, grayish sandstones, and arenaceous shales, very much variegated. 1st. Sandstones, fifty feet; 2d. Variegated arenaceous shaley clays, sixty feet; 3d. A curious wall of sandstone which forms a conspicuous point by turning the current of the creek at a right angle and running across, in a nearly vertical position, but having the natural dip northeast; dip 85°. This curious wall will always be noticed by travelers. It passes up gradually into the variegated sandy shales or laminated sandstones that form No. 12.

12. Among these laminated sandstones is a sort of silicious mud layer that is filled with the casts of a species of *Mytilus*, which leads me to suspect them to be jurassic. There is also a layer filled with fragments of fossils—a saurian tooth, &c. The beds continue with a reddish tinge varying from a greenish brown to a dull purplish tint, with every degree of texture. Some of the layers of laminated sandstone are a light ashen gray, some of nodular and pebbly sandstone, also with a tendency to lamination—300 feet.

13. A rather massive gray sandstone, some portions amygdaloidal or nodular, some fine grained and some slightly calcareous. Some of it is good for building purposes, flagging stones, &c. Two layers of ashen gray clay—first six feet, second three feet.

14. Very dull purplish clays, with some harder layers of sandstone, thin, of an ashen gray—30 feet.

15. Like bed 13, only more laminated, portions massive and fine; some layers a rusty yellow, with impressions of woody stems and trunks, not jointed but ribbed (!); passing into a dull purplish red massive sandstone, with a very irregular laminæ of deposition, some of it pebbly and nodular—200 feet to 300 feet.

16. Reddish laminated shale, with some greenish or ashen spots, some nodules, but slightly variegated with seams of fibrous gypsum following cleavage—300 feet.

17. Yellowish gray, rather fine grained, massive sandstone; portions of it with a reddish tinge; cleavage joints shown well—100 feet.

18. Reddish brown shales slightly gypsiferous—25 feet.

19. Massive sandstone, like 17; dip 75° to 80°—100 feet.

20. Very dull purplish drab, somewhat nodular, arenaceous clays with some hard layers of sandstone, mostly dark brown, and very variable in texture. This bed belongs to the lower cretaceous, or is a bed of passage—200 to 300 feet.

21. The sandstone "hog-back," regarded as lower cretaceous No. 1. A very conspicuous formation in this region. A portion of No. 1 stands quite vertical, while other portions incline from 60° to 80°. It is in part a coarse sandstone and fine aggregation of pebbles, passing up into a fine grained whitish sandstone, two hundred feet thick, passing to a series of alternate thin layers of dark laminated clay and mud sandstones, with all sorts of markings, indicative of shallow water, mud flats, &c. The dip of some of the layers passes a vertical at the top.

22. Then come the dark clays of No. 2, slightly arenaceous at first, passing up into black shales, then into the blue marly limestone with an abundance of *Inoceramus*. Some of the layers of blue limestone have

passed a vertical position, 30° to 40°. No line of demarcation can be found between; he divisions of the cretaceous. They all pass into each other imperceptibly.

The cretaceous beds are well shown, No. 2 continuing up into blue marly limestone, which may be regarded as No. 3; this passing up into the dark shales of No. 4, which gradually passes up into a rusty yellow clay with numerous calcareous concretions with *Ostrea*, *Baculites*, &c. This bed contains calcareous sandstones filled with a small *Turritella* and bivalves. The cretaceous rocks of this region are best divided into upper and lower cretaceous. These beds become suddenly horizontal in the plains, but the conformity is complete. The conformity of the entire series of the sedimentary beds is more perfect than I have seen it at any other locality in the West. Here, for the first time, I notice the two sets of red beds mentioned by Dr. Newberry, in his report of the Colorado River. They are well defined. The cretaceous beds are well marked. In the section, from beds 11 to 19 inclusive, I am inclined to regard as jurassic; the second series of reddish beds, as triassic; then some reddish permian (?) sandstones, passing down into the carboniferous.

Above the springs there is an extensive series of gneissoid rocks, inclining northwest. The changed and unchanged beds do not conform. These gneissic rocks vary much in texture and color. The dominant constituents are reddish feldspar and quartz, but there are thick beds of the banded gneiss. For about two miles up the Gallinas Creek, above the springs, these rocks rise up in grand mountain masses, nearly vertical, and then for ten miles or more we find the limestones, sandstones, and shales of the carboniferous, resting in a nearly horizontal position over the vertical edges of the gneiss. About four miles above the springs I found two distinct species of lepidodendron in sandstone, one of them twelve feet long. They leave a cast in the sandstone perfectly round. Still further up the creek we see the limestone resting directly on the gneiss for half a mile. Usually these beds are so covered by debris that they are obscured. As we pass up the creek the carboniferous beds come down to the water's edge. Three beds of limestone, from ten to thirty feet thick, are exposed on the sides of the hill.

About eight miles above the springs the valley expands out, and the gneissic and basaltic rocks form the lower mountain ridges. At the head of the valley there is a very striking basaltic mountain, with nearly perpendicular sides, which forms a land-mark in this region.

The hot springs are most beautifully located in the valley of Gallinas Creek, just as it emerges from the mountains on the south side. The springs are twenty or thirty in number, and some of them are quite large. They vary in temperature from 80° to 140°. The spring from which the water is taken for the bath is quite hot, at least 140°. The supply is very abundant, enough to meet the demand for all time to come. There is no deposit about the spring, and the water is as clear as crystal. It was analyzed by Mr. Frazer, and found to contain carbonate of soda, carbonate of potash, and chloride of sodium, the potash in excess.

It will be seen at once upon what its medicinal qualities depend. Every day in the week all the springs are occupied by women, in washing clothes. The water makes most excellent suds, and the ease with which the dirt is extracted from the clothes renders these springs great favorites. There is every facility for the proprietors to establish a place of resort for invalids and pleasure-seekers, when there shall be a sufficient demand.

West of the town of Vegas there is an almost vertical wall of cretaceous sandstone, running nearly north and south. Passing south along the east side of this wall about five miles below Vegas, we enter the hills through a gorge in this sandstone, called the Puerto cito del Padre. West of this we can see the complete series of the sedimentary beds in the form of upheaved ridges, rising like steps to the main mountain range. Our course to Santa Fé was nearly south, through a very rugged country; the formations thrown up into lofty ridges. In a few miles we came out into open valleys quite broad, and nearly all the beds are older than the cretaceous, and nearly horizontal in their position, and the valleys have been carved out of the jurassic and triassic beds; the very singular castellated hills on the left looking much like mesas, capped with sandstone, probably cretaceous in part, showing the red beds beneath. Sometimes the entire series of red beds are clearly shown, and even the carboniferous limestones are well exposed, but the Gallinas section is so complete that I need not repeat it here. On the San José Creek and the Rio Pecos are some fine exposures of the triassic rocks, showing their peculiar features and their variable texture. The prevailing color of the upper series of variegated beds is brick red, and that of the lower, or supposed triassic, is dull purplish. Close to the village of San José the beds are all nearly horizontal. The high hills around retain their mesa-like form. Nearly all the way to Apache Cañon, on the crossing of the mountain, the road runs along a valley with a lofty ridge or "hog-back" on one (the east) side, which forms a sort of outer wall and a conspicuous feature. The upper series of red beds are well shown, and a portion of the second series, but the bed of sandstone which caps the ridge, I am inclined to regard as a part of the jurassic group. At any rate I have not been accustomed to include these yellowish-gray, fine-grained sandstones among the cretaceous beds. The ridge of sandstone which forms the outer wall at Vegas must still continue outside of this ridge.

In the lower red series are beds of gypsum. I saw gypsum at a number of localities on our route. At Vegas, it is used for building purposes.

I was informed that a coal mine had been found near Tecalope, and that the coal had been used for blacksmithing, but I saw no rocks that could possibly contain coal, on my route, and think that it was a mistake. A copper mine has been wrought with some success near San Miguel in the triassic beds. I did not examine it. Near Pecos Creek, all the rocks seem to be in contact, from the light-colored sandstones, that cap the mesa, to the carboniferous. All the beds dip a little from the main range, and this inclination increases as we approach the mountains. The width of the belt of upheaved sedimentary rocks, from Vegas to the southern point of the mountains north of Gallisteo Creek, must average twenty to thirty miles, and the opportunities for studying them, in their order of sequence, is excellent.

At Payaritos Springs station, there is a splendid exhibition of the different groups of strata, as we have seen them since leaving Vegas. The light-gray sandstones and first and second series of variegated beds are all shown in their order for six hundred to eight hundred feet.

About six miles north of the old Pecos church, there is a bed of compact reddish limestone, full of fossils, which I am inclined to regard as permian, containing fossils *Myalina*, *Mytilus*, *Pleurophorus*, and crinoidal stems. This limestone belongs to the lowest portion of the second series of red beds. I would just remark here, that I am inclined to the belief that in the mesa, which looks so conspicuous on our left, on the road to Santa Fé, we have the first series of variegated beds, or jurassic, including the fine-grained sandstones that cap them; and the second series, triassic;

and that the remaining sedimentary beds are composed of carboniferous and possibly some permian exists in this region.

The carboniferous fossils are unmistakable, and I think I have collected some permian-like forms, and I suspect that my collections will furnish the evidence of the age of the upper series of red beds as jurassic. The dip of all these beds is slight, not more than five to eight degrees. From the old Pecos church to the Apache Cañon, we pass over the beds of the lower red series, mostly consisting of sandstones and fine pudding-stones. But in Apache Cañon, we have a good exhibition of the fine-grained, light-colored sandstones, which have capped the mesas on our left for thirty miles or more. So far as I could determine, this sandstone does not belong to the lower cretaceous, but the true cretaceous ridge is still further east. The inclination of the strata in Apache Cañon is sometimes twenty-five degrees southward.

As we commenced the ascent of the mountains towards Santa Fé, the surface is covered with a remarkable conglomerate, a paste of sand and clay holding fast unworn masses of reddish granite. I think that this is a modern formation, and underneath it we find the dull purplish-brown sandstones. I did not notice the carboniferous limestones on the east side of the range, but do not doubt that they exist high up on the mountain sides. From the summits of the mountains we can look far southward. All the ridges of upheaval continue southward along the flanks of the mountains, and soon run out in the plain, and the mountains slope down to the prairie about twenty miles south of Santa Fé.

About thirteen miles before reaching Santa Fé, we come to the gneissoid rocks, and they continue nearly to that place. They seem to dip with the sedimentary rocks on each side, only at a higher angle. This mountain forms a regular anticlinal. On the flanks of the mountains, (west side,) there is quite a thick deposit of yellow and light flesh-colored marls and sands extending westward toward the head of the Rio Grande, and beyond. The mountains themselves seem to be quite peculiar, in being composed of an aggregate of cone-like peaks of very variable heights. They seem to be entirely composed of gneissoid rocks.

CHAPTER VI.

FROM SANTA FÉ TO PLACIERE MOUNTAINS AND RETURN.

From Santa Fé to the banks of the Gallisteo Creek, eighteen miles, we pass over the recent marls and sands which seem to occupy the greater portion of the valley of the Rio Grande, above and below Santa Fé, which I have called Santa Fé marls. These are mostly of a light cream-color, sometimes rusty yellow, and sometimes yellowish white, with layers of sandstones, varying in texture from a very fine aggregate of quartz to a moderately coarse pudding-stone. These marls and sands weather into unique forms north of Santa Fé, like the "bad lands" or "Mauvais Terres" of Dakota. As we descend the hill into the valley of the Gallisteo Creek, we have a wonderful exhibition of the variegated sands and sandstones, which at first appear like the upper series of red beds on the east side of the mountains, but which I at once suspected were new to me in this region. Descending the Gallisteo, to the west or lower end of the Cerillos, we find the full series of the cretaceous beds, with *Ostrea congesta*, *O. larva*, *Inoceramus*—several species, and fragments of fish remains. Extending east and west along the south side of the

Cerillos is a high wall-like dike, and dipping southward from this, from the Placiere Mountain, is a great thickness of the cretaceous shales No. 2, passing up into laminated arenaceous shales, with fossils, then the dark shales of No. 4, apparently. The cretaceous beds incline thirty degrees to fifty degrees. Inclining at a less angle, a series of coal strata reveal their upturned edges, conforming perfectly to the cretaceous beds. Passing up the Gallisteo, eastward, we observed the variegated sands and sandstones, rising above the coal strata, and concealing them on the northeast and east flanks of the Placiere Mountain, inclining at all angles from five degrees to fifty degrees. These sandstones are of varied texture, from a fine aggregate of quartz particles to a rather coarse pudding-stone. In some of the beds there are irregular layers, of a dull, rusty brown, concretionary arenaceous limestone, in which I searched in vain for fossils. One of the most peculiar features of these beds, and one which I have never seen in any group before, is the great variety of colors, from a light reddish tint to a deep brick red, sometimes dull purplish light, and very deep yellow, white, brown, drab, &c. The only fossils I could find were enormous silicified trunks of trees. One of them was so perfect that it looked much like a recent one, with a cavity running through it ten inches in diameter. I have named these beds the Gallisteo sand group, as they are confined, so far as I know at present, to the valley of the Gallisteo, although they pass under the Santa Fé marls, and the northern limit is concealed from view. Near the road is a small dike, apparently thrust up between beds of sandstone, and inclining with them. East of the Cerillos, up the Gallisteo, among the upper beds of that group, are several larger dikes, and the basaltic rocks are poured over the recent tertiary beds. One of the dikes can be seen a long distance, looking like a ridge of upheaval, extending a little north of east, far across the plain towards the south end of the Santa Fé Mountains. The Cerillos are merely a dike, or a series of dikes, forming a small independent range of mountains composed entirely of eruptive rocks. On the south and west side, the cretaceous beds flank them closely, while on the east and northeast side the Santa Fé marls jut up against them. Occasionally, on the east side, a little stream will cut through the marls, revealing the sandstones of the Gallisteo group.

The outcroppings of coal on the northwest side of the Placiere Mountains are of great interest. They were first exposed in the center of the little branches that run into the Gallisteo. The first one is about five miles south of the Gallisteo. The coal is in the natural condition. The following section of the strata was taken ascending:

1. Laminated clay, with thin seams of sand passing up into carbonaceous clay, as a floor for coal.
2. Coal very compact. The cleavage lines are, in a few instances, filled with clay—5 to 6 feet.
3. Drab clay, indurated, 15 to 29 feet.
4. Ferruginous sandstone, passing up into a light grayish sandstone—30 to 50 feet.

The lower part of this bed is full of deciduous leaves. The debris is so great that the real character of the beds is somewhat obscured. The impressions of leaves, appear to belong to the genera *Magnolia*, *Platanus*, *Salix*, and others, some of which appear to be identical with those found at the Raton Mountains. Imperfect specimens of a palm were found. The mine is opened on each side of the dry creek, and the dip is the same—about ten degrees. As in all the rocks of the country, there are in the coal two sets of cleavage lines, at right angles to the stratification. In the valley of another branch of the Gallisteo, there is a dike two feet wide with the strike a little east of south. The clays on each side are metamorphosed into slates.

At another locality there is a bed of coal, which has been changed by an enormous dike into anthracite. Section 1st, clay-slate; 2d, two and a half to three feet anthracite; 3d, fourteen to eighteen inches of clay; 4th, fourteen inches to two feet of anthracite; 5th, clay shale, passing up into alternate layers of sandstone and clay, ten feet; 6th, dark sandstone. The dip of all the beds is fourteen degrees east. They are overlaid by a thick bed of columnar basalt. The dike that covers the coal bed trends about north and south, or a little east of south.

The influences of the Cerillos on the north side of the Gallisteo, and the Placiere Mountains on the south, has produced a beautiful synclinal, while the Cerillos form a sort of imperfect quaquaversal. The beds dip from two sides of this small range at least, and the indications in the channels of the little streams are, that the sandstones of the Gallisteo group dip from a third side, but are now mostly concealed. We have, therefore, in the valley of the Rio Grande, if my investigations are correct, three groups of tertiary beds of different ages. 1st. The coal strata, with abundant impressions of deciduous leaves, lying above well-marked cretaceous beds. 2d. The Gallisteo sand group, which plainly overlies the coal strata, but inclines equally with and conforms to them. 3d. The Santa Fé marls, which are of much later date than either of the other two, and rest unconformably upon the Gallisteo group, and never incline more than five or ten degrees.

Although the coal beds lie high up on the sides of the Placiere Mountains, I am inclined to the belief that some portions of the cretaceous strata, and possibly even older rocks, are revealed on the sides of the gneissic nucleus. This mountain seems to be penetrated with dikes, which have given a dark somber hue to all the rocks.

The mountain itself is very rich in minerals, as gold and iron ore. The Ortiz mine is the most celebrated, although a number of lodes have been opened. Colonel Anderson, formerly of the United States Army, is superintendent of the mining interests of this region, and he has already erected a forty-stamp mill, the most substantial one I have seen in the West. The Ortiz lode is a very irregular one. It expands sometimes twelve feet or more, and then nearly closes up. It has yielded very rich ore at times; mingled with this ore are fluorspar, calcspar, crystallized quartz, blue and green carbonates of copper, a little native copper, and the sulphurets of iron and copper—the latter predominates in the ore.

The Brehm lode has a strike about northeast and southwest, and by it I suspect the dip of the gneissic rocks, on the north side of the mountains, to be about northwest. The width of the lode is about three feet, the inclination of the vein southeast forty-five degrees.

The Placiere Mountain seems to be rich in gold, but the want of water may prevent the mines from being wrought with great profit. The surface of the country is literally covered with placer diggings, where the drift gold has been taken out by the Mexicans in old times by melting snow. Magnetic iron ore seems to be abundant, and in the clays connected with the coal beds there is the largest supply of excellent brown iron ore, so that the time is not far distant when iron works of great value may be erected in this region.

CHAPTER VII.

FROM SANTA FÉ TO TAOS.

On the western flank of the Santa Fé Mountains, near Santa Fé, I found the foot-hills, which are exposed by the wearing away of the marls, to be composed of carboniferous beds. These beds of limestone rest directly on the granite, and are associated with gray and reddish shales and some beds of sandstone, the whole dipping west at an angle of thirty-five to forty-five degrees. The limestones were charged with fossils, as many and as well preserved as I have seen them at any locality east or west. In many places these beds of limestone are carried high up on the granite hills; sometimes dipping toward the mountains as if a portion of an anticlinal. The metamorphic rocks are gneissoid at first on the flanks, but gradually become massive granites toward the main axis of the range. In a small creek, which leads down from the mountains, I saw immense masses of granite breccia, mostly angular fragments of gneiss or red feldspar, with some rounded masses cemented with a granite paste. The limestones about Santa Fé are converted into excellent lime. The foundations of the jail and court-house are made of it. The fossils are very numerous, both in individuals and species. Dr. Newberry has given a list of them. I found several species of *Productus*, *Sprifera subtilita*, and many others. These limestones do not seem to extend far along the sides of the mountains. From Santa Fé to Embudo Creek, and mostly even to Taos, the Santa Fé marls cover the country. On the east side of the Rio Grande I did not observe a single dike, from the Cerillos to the mouth of the Chama Creek. North of that the melted material has been poured over the marl so as to form broad mesas. On the west side there are numerous outbursts of igneous matter. These Santa Fé marls reach a great thickness north of Santa Fé, in the Rio Grande Valley, from one thousand two hundred to one thousand five hundred feet, and have a tendency to weather into similar monumental and castellated forms, as in the "Bad Lands." The upper portions are yellow and cream colored sandstones, sands, and marls. Lower down are some gray coarse sand beds with layers of sandstone. All these marls dip from the range westward three to five degrees. The Rio Grande wears its way through these marls with a bottom about two miles wide. On the west side are distinct terraces with the summits planed off smoothly like mesas. The first one is eighty feet above the river; the second one, two hundred feet. These marls extend all the way between the margins of the Santa Fé Mountains on the east side and the Jemez Range on the west. Most of the Chama Hills, and I think the entire hills, are composed of it. At the junction of the Chama Creek with the Rio Grande, a point comes down between the two rivers which is covered with basalt. This continues into the San Luis Valley nearly to Fort Garland. Near the mountains the hills are covered extensively with drift, and sometimes they are composed largely of boulders and marl or sand. The country is full of the dry beds of creeks or *arroyas*, as they are called. All these carve their valleys out of these marl beds. As we go northward near the mountains, the rounded boulders become more and more numerous, but near the Rio Grande, where they have all disappeared, the source of this great thickness of sediment is apparent.

The Ojo Sarca Creek rises in the Sante Fé Mountains and flows into the Rio Grande, forming a valley which is remarkable for its ruggedness. The marl beds are nearly horizontal and the harder layers of sand project out of the sides of the bluff hills like steps for four hundred to six hundred feet in height. I know of no finer exhibition of these marls in their thickness, or their architectural style of weathering. On the north side of the creek, the granites project up through the marls.

The mountains near the source of the Rio Trampas are very lofty, with some high peaks which are rounded with dome-like tops, one of which is called "Old Baldy" from its bare summit. Where the foot-hills are denuded of the drift or the marl, the red granites are exposed. Along the base of the mountains, especially in the valley of the Peñasco there is a great thickness of very coarse conglomerate resting upon the granite horizontally. It undoubtedly is of the same age as the marl beds. In the valley of the Peñasco there is a vast quantity of worn boulders, scattered everywhere, similar to the valley of Boulder Creek in Colorado. These worn rocks are of large size next to the mountain, but diminish the further they recede to the westward.

The valley of the Rio Grande is already settled by Mexicans wherever there is an available spot. Nearly all the land that can be irrigated is cultivated by them, and good crops are raised even with their rude style of cultivation.

CHAPTER VIII.

FROM TAOS TO FORT GARLAND.

The valley in which Taos is situated may be said to be formed by a notch or bend in the mountain range. On the southwest is the Pickaris Range, with a strike nearly northeast and southwest. The next range east of this trends about north and south, and branching off from this, north of the Taos Valley, are the Pueblo Mountains, Dos Mountains, and the Rio Colorado Mountains, all with a strike nearly northwest and southeast. The course of the Rio Grande is nearly south, and on each side of Taos the small ranges of mountains run out near the river. The notch or bow in this group of mountain ranges affords a fine illustration of the method of flexure in the mountain ranges.

The Taos Valley is about eighteen miles in extent, from east to west, and about sixteen miles from north to south. It is thickly settled by Mexicans, and every available spot of ground is taken up.

The valley proper is scooped out of the Santa Fé marls, which must at one time have prevailed extensively, as in the country north of Santa Fé, but the surface has been smoothed off, so that nowhere are the marls conspicuous; still they can be seen all along the base of the mountains bordering the valley where portions of the recent deposits lie high on the mountain side. No sedimentary rocks of older date are seen, and the Santa Fé marls rest directly on the metamorphic rocks.

It is plain that the regular metamorphic rocks prevail in these mountain ranges, but mingling with them in various localites are igneous outbursts, which have somewhat tinged the gneissoid rocks. A little south of Taos River we find beds of beautiful porphyritic breccia, which is very compact, and is employed for building purposes. Westward, toward the Rio Grande, it is probable that the broad level plain is underlaid with a sheet of basalt, for the Rio Grande itself runs through a very deep cañon of this material for sixty-five miles, from La Joya to the crossing of the road to Conijos in the San Luis Park. In all this distance there is but one crossing for teams, and three others for persons on foot, and there the passage is made with great difficulty. Far distant, to the west of the Rio Grande, are numerous isolated mountains showing the igneous protrusions.

Taos Valley, therefore, forms a sort of half circle, and the mountains which surround it, of which there appear to be ten or twelve distinct ranges, are expansions of the main range. It is near this expansion that the Moreno mines are situated, which have already proved unusually rich, and will probably continue to yield large returns of gold for many years to come.

On the sides of the mountains immediately opposite the Morena Valley, north of Taos, are located the mines of the Arroyo Hondo Mining and Ditching Company, of which Mr. Lucian Stewart, of Taos, is the superintendent. Mr. Stewart has already erected a twenty-stamp mill on the San Antonio Creek, and the supply of water is so great that if the mines turn out to be rich in gold, the enterprise will prove a complete success.

About twenty lodes have been prospected with more or less encouragement, and some of them look well. In most instances the country rock has a greenish ashen tinge, doubtless due to the influence of heat from the igneous rocks. The lodes are not very well defined; one lode has a strike a little west of north. It contains carbonate of copper, sulphurets of copper and iron. It was first prospected for silver, but turned out to be richer in gold. The cleavage walls are lined with sulphate of lime. The gangue rock is mostly feldspar and quartz highly ferruginous.

The main lode of the company is situated about half way up the south side of the mountain. Dip of vein, thirty-five degrees, strike nearly east and west, inclining about south. The country rock is mostly quartz, quite hard, while the seam, which is pretty well defined, is rotten quartz. It is eight to twelve inches wide, and is called the "pay streak," although the neighboring rock pays well. There may be a very wide crevice here of which the walls have not been discovered. The cleavage lines are well shown, and are of two kinds, one set dipping south thirty-five degrees parallel with the ore streak, and the other inclining north twenty degrees. The principal lines of cleavage contain the rich ore. The dip of the country rock is plainly south or southeast at a very high angle. A tunnel has been excavated into the side of the mountain five feet in diameter, and one hundred and eighty feet deep, two hundred feet below discovery point.

All along the sides of the mountains are quite thick deposits of recent material, as clays, sands, and marls, and at one locality, while digging a ditch, Mr. Stewart discovered a thick bed of aluminous clay which contained much gold, but it was found to be so difficult to extract it that the placer was abandoned. The sides of the mountains everywhere are covered with "diggings," where the Mexicans in former times washed the loose drift with water, obtained by melting the snows.

These mountains are composed largely of gray granite, and the reddish feldspar is not much seen. Each one of these ranges seems to afford a good example of an anticlinal axis, the sides being shown by the shape of their slopes, which are very seldom symmetrical, one side of the anticlinal being much more prominent than the other.

From Taos to Rio Colorado the foot-hills of the mountains are covered with *piñon*, with a few larger pines which would make excellent timber. Indeed, I am inclined to the opinion that the basaltic mesas are the natural habitats of the *piñon*, which is a low scrubby tree, fit only for fuel, while the larger species of pine and spruce are found growing on the metamorphic rocks.

As we approach the Rio Colorado the outbursts of basaltic material increase. The Rio Grande and its branches, before they join the larger stream, show vast cañons with nearly perpendicular sides, and the peculiar dark somber color of the rocks adds to the gloomy picturesqueness of the scenery. On each side of the valleys of the little streams as they issue from the mountains are high terraces one hundred to one hundred and fifty feet high, which are here more conspicuous than usual. These are composed of the Santa Fé marls and are not unfrequently covered with a thick bed of basalt.

The broad intermediate space between the range of mountains which form the east side of the valley of the Rio Grande and the Sierra Madre—a main range of the Rocky Mountains, which gives origin to the waters of the Pacific streams—is covered with rounded hills, detached ranges, &c., all of which are basaltic. The two rounded hills, which are very marked, situated nearly opposite each other on different sides of the Rio Grande, Cerro de las Utas and Cerro San Antonio, are, it seems to me, old craters or fissures, out of which issued the melted material which overflowed the sides, and time has weathered the whole mass into its present beautiful rounded form. At this time they look like gigantic mammæ.

I am inclined to regard the valley of the Rio Grande as one great crater, including within its rim a vast number of smaller craters and dikes, out of which has been poured at some time a continuous sheet or mass of melted material. All the valleys, small and great, give evidence that they have been worn out of this vast mesa. The Rio Grande, from its source in the San Juan Mountains to Albuquerque, flows along its banks through basaltic rocks to a greater or less extent, and as we go northward from it they disappear in part.

By glancing at a map it will be seen at once that the valley of the Rio Grande belongs to the eastern side or Atlantic slope of the "great divide," and that the ranges of mountains, on the east side of the valley of the Rio Grande, which run out into the plains near Santa Fé, are a series of fragments which seem to have broken off from the main rocky range north of the South Park, and are prolonged southward in a more or less broken condition for over four hundred miles. It will also be seen from the map that the parting line or divide flexes over to the west, with a great bend above Middle and South Parks. Now between the base of this mountain prolongation on the east, and the Sierra Madre or western divide on the west, from the head of the San Luis Valley to Gallisteo Creek, at least, an area of over two hundred miles from north to south, and one hundred to one hundred and fifty miles from east to west, is occupied mostly by but two classes of rocks, the basaltic and the modern tertiary or Santa Fé marls. These recent marls are very conspicuous about Santa Fé and north of that point to the Pickaris Mountains, but north of that point they are not largely developed, though at one time they must have reached a great thickness, but have been removed by erosion. The valley of the Rio Grande, from Santa Fé to Taos, has the appearance of an immature region, much like that of the "Bad Lands," or the tertiary deposits of White and Niobrara rivers on the Missouri. But above and north of Taos the wearing and smoothing process has been applied so that there is a mature appearance of the country, like that of Eastern Nebraska or Kansas. Still all along the foot of the mountains below Costilla, underneath the mesa which extends from below Costilla to the Sierra Blanca Range, fifty miles, these marls can be seen in places. At Culebra, the Rio Culebra cuts through the mesa, forming a sort of gorge nearly half a mile in width. On the sides of the mesa these marls are most clearly seen underneath a heavy bed of basalt. Along the little branches of the Rio Trenchara, as the Rio de las Utas and the Sangre de

Christo and on the south side of the Sierra Blanca Range, are prominent terrace-like hills which are composed of yellowish-brown marls and sands. On the south side of the Sierra Blanca, they jut up high and close on the mountain slope. These marls are only remnants of large deposits which once existed here, and spread out uniformly all over the valley.

That there are mines of gold and other precious metals, as well as iron and copper, in the mountains, along the eastern side of the San Luis Valley, has long been known. Specimens of copper, indicating mines of considerable richness, have been brought from the sources of the Costilla and iron ores are scattered all over the valley of the Rio de las Utas. In the foot-hills of the mountains are fragments of magnetic iron ore, much like that in the valley of the Chugwater Creek. Stray masses have been traced up the mountain sides for about five miles, where a "blow-out" or an immense mountain mass has been discovered. This iron occurs in the gneissoid rocks, or what is called the Laurentian group, to which group, I believe, all the gneissic and perhaps the entire mass of metamorphic rocks of the Rocky Mountain system belong. I have assumed the position, in all my investigations, that there are but two classes of changed rocks in the West, viz, igneous and metamorphic, and that the oldest granites which form the nuclei of the loftiest mountain ranges were once aqueous rocks, deposited in the same manner as the limestones or sandstones of our most modern formations. It is on this ground that I have so often used the terms "changed" and "unchanged" rocks. By igneous rocks, I always mean those only that I regard as having once been in a fluid state, and may or may not have been protruded so as to reach the surface. I also assume that these igneous rocks in the West may have been thrust up at different geological periods, or at different times during the same epoch.

The gold mines near the Sangre de Christo Pass are the most important that have been discovered in the San Luis Valley. From some notes kindly furnished me by Dr. McClellan, United States Army, surgeon of the post at Fort Garland, the history of these mines appears to be as follows:

During the gold excitement in the San Juan Mountains, west of the Rio Grande del Norte, in 1862, a large number of miners, or, as they were called in those days, "pilgrims," crossed the Sangre de Christo Pass, and camped for rest after a long journey from Idaho, Montana, and Northern Colorado, on Placiere Creek, one of the main tributaries of the Rio del Sangre de Christo. Learning from some passing Mexicans, that in the olden time their people were accustomed to pack dirt from some of the cañons of the mountains to the Placiere Creek, to wash out the gold, they went to work and prospected the gulch of the Grayback Creek. The San Juan excitement was, however, so strong that they started to continue their journey the winter of the same year, many of whom returned in a starving condition, and went to work in this gulch with good results.

In 1865 and 1866, Kit Carson with a party prospected this region for placer diggings, and took up many valuable claims. The gold taken out by sluicing is very valuable and of a pure yellow color, and is what is called "wire and scale" gold. It usually sells for about $19 per ounce in gold, much more than the Morena gold or any other in this country. A valuable lode with a well-defined crevice has been uncovered, but little or no work has been done on it. In the mountains at the sources of the Rio Seco, on the west side of Culebra Peak, some lodes have been found which appear favorable. Most of the foot-hills are covered with beds of yellow marl inclining slightly. These foot-hills seem to be smoothed off and are covered with a thick deposit of debris. In the little valleys of the mountains the gneissic rocks are exposed, and about twenty lodes have been examined to some extent; the crevice matter on the surface is entirely rotten quartz. These lodes have in most cases well-defined walls, varying from three to six feet in width, and a strike about northeast and southwest. Should these mines turn out to be rich in gold, the ease with which they can be worked will render them very celebrated.

From a point not more than twenty miles north of Santa Fé, to the Sangre de Christo Pass, I was unable to discover any of the older sedimentary beds on the western side of the mountains. Sometimes among the drift boulders, which were very extensive everywhere, a few masses of limestone would be found which were evidently carboniferous. In Taos Valley slightly worn masses of limestone were found, with well-defined carboniferous fossils. This would seem to indicate that these rocks once existed all along the mountains, even if they cannot be found at this time. I have no doubt that all the sedimentary formations which are found on the eastern margins of the mountains once extended uninterruptedly across the Rio Grande Valley, and some portions may now exist deep beneath the basalt and Santa Fé marls.

Near the Rio Colorado, the lower ridges or foot-hills of the mountains exhibit the influence of the igneous rocks to a greater extent than southward, and continue to do so to the Sierra Blanca. Near the point from which the Rio Colorado emerges from the mountains, the rocks are a bright brick-red over a small area, and I mistook them for remnants of the triassic. A closer examination showed me that high up on the sides of the mountains a great thickness of the recent marls, sands, and clays, have been so changed by contact with the igneous rocks, that they now present that peculiar brick-red and variegated appearance which is noticed for several miles.

At Costilla the main range seems to bend abruptly to the eastward, and a portion of the lower ridges on the western sides of the mountains south of Costilla passes off without interruption in a long basaltic mesa, which extends nearly to Fort Garland. East of this mesa are the "vegas" or meadows, which have been carved out of the mesa between it and the foot of the mountains and form a portion of the valleys of the Costilla and Culebra Rivers. North of Culebra the basaltic mesas commence again close to the base of the mountains, and continue quite largely developed up to the Sierra Blanca Range. These mesas are capped with a heavy bed of basalt, which always seems to incline eastward toward the mountains at least from three to five degrees, and sometimes much more.

On the east side, close to the Rio Grande, near the entrances of the Trenchera and Culebra Rivers, are a great number of ridges and conical peaks or hills, called "Cerillos," all of them basaltic. On the opposite side of the Rio Grande these basaltic hills are very abundant, and occupy most of the country. Just north of the Trenchera this range of mountains seems to bend abruptly back to the westward in the form of the Sierra Blanca Mountains, which have a trend nearly east and west. There is therefore a quadrangular space inclosed on three sides by mountains—the Costilla on the south side, about fifteen or twenty miles; the principal range on the east, about sixty miles, and the Sierra Blanca on the north, about fifteen or twenty miles. The main range continues northward, bending slightly westward, until it joins the Sierra Madre at the Poncho Pass. The Sierra Blanca is the grandest and most picturesque range in Southern Colorado. It is apparently basaltic and is, as I suppose, a gigantic dike. I regard the Spanish Peaks as an enormous dike. In nearly all cases the strike of the axes of these dikes is nearly northeast and southwest, while for the most part the axes of the granitic ranges trend about northwest and southeast. It is my opinion also that the elevation of the basaltic range was an event subsequent to that of the granitic, for in all cases that I have ever examined the igneous rocks are poured out over the granites, and in some cases concealing them entirely over large areas. Many of the loftiest peaks in the granitic ranges are basaltic. The basaltic axis never passes through the granitic, as is shown by the Spanish peaks on the east, and the Sierra Blanca on the west. Each one stops abruptly as it comes up against the principal granitic axis.

As I have before observed, no unchanged sedimentary rocks of older date than the Santa Fé marls were noticed along the western side of the main range north of Taos, until we come to the Sangre de Christo Pass. About the sources of the Costilla or Culebra Rivers there may be some remnants, but none have been seen after a pretty close examination. At the very summit of the pass is a series of reddish sandstones and shales nearly vertical but inclining westward. From the summit of the pass we descend the beautiful valley of the Sangre de Christo Creek in a southwesterly direction. The sandstone extends for four or five miles, and is of every variety of texture, from a very fine compact silicious rock to a moderately coarse pudding-stone. In some of these sandstones are indistinct vegetable impressions, some of which can be recognized as fragments of *Calamites*. Further down we come to a series of limestones and sandstones, with some calcareous sandstones, having thin beds or partings of shale. These alternate limestones and sandstones extend for about five miles, and then comes a belt of five miles of gneissoid granites. Near the junction of the limestones with the granites there is a bed of limestone filled with fossils, *Productus*, several species, *Spirifer subtilita*, *Rhynconella rockymontana*, *Spirifer lineatus*, and numerous corals and *crinoidal* stems. Although it is possible that there are here rocks of older date than carboniferous, yet from the fact that all along the eastern side of the mountains the carboniferous limestones have been found resting upon the granites, I have inferred that there are no sedimentary rocks of older date in this region. At first the Sangre de Christo Creek passes through a monoclinal rift for nearly ten miles, then it cuts through ridges of limestone, bed after bed. The real dip of all these beds is northeast while the apparent dip is southwest, as if the granites were more modern than the limestones which are above them. The belt of granites is about five miles wide, and thence to Fort Garland, which is ten miles, are igneous rocks. Eight miles east of Fort Garland are some high ridges of basalt that dip east about eight degrees, and have a trend north and south, and from the abrupt western face from four to six distinct beds of igneous rocks can be seen. The cause of the inclination of the basaltic beds is not clear, though it may have originated in the Sierra Blanca. We were much indebted for many favors and information to Dr. E. McClelland, surgeon, and to Colonel Hart, commander of Fort Garland.

CHAPTER IX.

FROM FORT GARLAND TO SOUTH PARK.

The Rio Grande del Norte River rises in the Park of the Animas, flows east about one hundred and fifty miles to the San Luis Valley, then bends abruptly south through the middle of the San Luis Valley. The northern portion of the valley is called the San Luis Park proper. This northern portion, above the bow of the Rio Grande, is about sixty miles in length, and has an average width of fifteen to twenty miles. About the center of this park is a singular depression, about ten miles wide and thirty miles long; it looks like one vast thicket of "grease wood," *Sarcobatus vermicularis*, and other chenopiaceous shrubs. Into it flow some twelve or fifteen good sized streams, and yet there is no known outlet, neither is there any large body of water visible. It seems to be one vast swamp or bog, with a few small lakes, one of which is said to be three miles in length. Although entirely disconnected from any other water system the little streams are full of trout.

On the south side of the Sierra Blanca the foot-hills are composed of the light-colored marls, and on the west side of the mountain, and near Mosca Pass, are the sand hills, which are composed of the loose materials of this formation.

Here also is another conspicuous remnant of it left after erosion. On the west side, just below Sawatch Creek, and in the Rincon, are some rather high hills of this marl at the base of the mountains. The materials thrown out of the excavations of prairie dogs show that the valley is entirely underlaid with it. I am convinced therefore that this fresh-water deposit occupied the whole of this valley from Poncho Pass to the mouth of Gallisteo Creek, and how much further southward I cannot tell, but there is evidence that it extends, either continuously or with interruptions, through New Mexico, and even further.

From Fort Garland to the Poncho Pass no sedimentary rocks of older date than the marls are seen along the margins of the mountains on either side until we reach Kerber's ranche, about ten miles below the summit of the pass. On the west side of the valley, on the foot-hills, is a large thickness of carboniferous limestones, lifted high on the summits, and dipping east at an angle of fifty degrees. This limestone continues only a few miles, and is another of the remnants that are left of the sedimentary rocks among the mountains.

Commencing at Fort Garland, the range of mountains that wall in the San Luis Park on the east side is grand in its proportions. From the Sierra Blanca nearly to the Poncho Pass it appears to be purely eruptive, and to be composed of a series of ranges or axes trending nearly northeast and southwest. At the northern end the eruptive portion ceases, and the lower metamorphic mountains flex around so as to trend northwest and southeast. On the west side, the mountains are far less lofty, but they seem to form a nucleus of metamorphic rocks, with a vast number of dikes, from which the basalt has poured over nearly the entire region. All the foot-hills south of the Sawatch are composed of eruptive rocks, but north of that point the gneissic rocks are seen. This range of mountains seems to be made up of a number of smaller ranges, with a general trend northwest and southeast. It would seem that where a range of mountains is purely eruptive the minor ranges trend northeast and southwest, but that where there is a metamorphic nucleus the eruptive materials follow the strike of the minor ranges.

At the summit of the pass the hills are grass-covered and the road excellent, but the nucleus of the mountains on the east side is metamorphic, with dikes of eruptive rocks everywhere. The little stream, the valley of which we descend, flows through a monoclinal rift or interval between the ridges of metamorphic rocks.

About two miles from the summit this little branch is joined by the main fork, and the whole continues to flow through a monoclinal valley until it empties into the South Arkansas. The main Poncho Creek rises in one of the loftiest peaks in Colorado. This peak has a large depression on the east side, which may once have formed a portion of the crater. At the junction of the forks commences one of the most remarkable examples of what appear to be igneous rocks I have ever seen in the West. On the east side of the creek we have the steep slopes, and on the west the projecting edges. We have here eight hundred to one thousand feet of eruptive rocks with a somber hue, but with a stratification as perfect as in any sedimentary rocks. It is composed of layers never over one to four inches in thickness, inclining south of west forty-five degrees. Some of the layers would make good flagging stones.

A little further down we come to the gneissic rocks, inclining northwest fifty to sixty degrees. Some of the black-banded gneiss has zigzag seams of feldspar and quartz running through them.

About three miles before reaching the Arkansas there is a curious junction of the massive red feldspathic granites, inclining northeast seventy degrees, with the dark-banded gneiss, inclining northwest twenty-five degrees. At the point of synclinal junction all is confusion; the two kinds of rocks are crushed together, and yet there is no break in the mountain. As we emerge from the pass to the South Arkansas we have the finest exhibition of banded gneiss I have seen in the West. The rocks are of various colors, red, yellow, white, and black, and the layers are quite thin, and their appearance is very picturesque. The general course of the Poncho Creek, from its source in the snow peak to the Arkansas, is north.

The gneiss is very varied in its texture; some of it contains garnets; some of it is very close feldspathic, micaceous, or whitish quartzose.

On the east side of Poncho Creek, about one hundred and fifty feet above the Arkansas, on the side of the mountain, is a hot spring surrounded with a large tufaceous deposit. There is also near the foot of the pass, on the side of the mountain, an extensive deposit of the yellowish marl, filled with water-worn boulders.

Between the South and North Arkansas there are some remarkable terraces or benches extending the whole breadth of the valley from mountain to mountain. On the north side of the South Arkansas are three terraces, beside the rounded hills near the base of the mountains, which rise in succession like steps.

The high eruptive range which seems to cross the South Arkansas, and to pass up along the west side of the North Arkansas, appears to be composed of a series of enormous dikes in a chain merging into each other, and having a strike about northeast and southwest. The general trend of the aggregate is about north and south.

On the west side of the Arkansas Valley the recent tertiary beds run up to and overlap the margins of the mountains. They are composed mostly of fine sands, arenaceous clays, and pudding-stones, cream-colored arenaceous clays, and rusty yellow marls, fine sand predominating. These beds weather into peculiar architectural forms, somewhat like the "Bad Lands" of Dakota. Indeed they are very nearly the same as the Santa Fé marls, and were doubtless cotemporaneous, and dip at the same angle, three to five degrees, a little west of north. The tops of the hills have all been planed down as if smoothed with a roller. I have called this group the Arkansas marls. They occupy the entire valley of the Arkansas. This valley is about forty miles in length, and on an average about five to ten miles in width. It might properly be called a park, for it is completely surrounded by mountains. On the west side is one of the grandest ranges of eruptive mountains on the continent. On the east side is also a lofty range with a metamorphic nucleus, but intersected everywhere with basaltic dikes. The first and lowest range runs parallel with it, and is sometimes cut through by it. It seems to be composed of massive feldspathic granite of igneous origin.

Near the mill, on a little branch just below the mouth of Trout Creek, there is a high rounded peak with a crater-formed depression at the summit which is grassed over, while all around the rim there is a fringe of pines. I am inclined to think it is an old volcano.

At the point where Chalk Creek emerges from the eruptive range, the sides of the cañon present a singular white chalky appearance. This seems to be due to the decomposition of the eruptive rocks, which appear to be true dolerite.

The drift evidences in this valley are very conspicuous. All along the Arkansas and in the valleys of the little branches are very thick beds of water-worn boulders of all sizes. The last of the eroding forces seems to have come from the range of mountains on the west side.

The granite on the east side of the river possesses, in a wonderful degree, the tendency to disintegrate by exfoliation. There is a kind of bedding which breaks the exfoliation or confines it. In these massive granites there are two sorts of cleavage besides the lines of bedding; one of these is usually vertical and has a strike northeast and southwest, and the other southeast and northwest, inclining twenty to forty degrees.

On the summit of the mountains is a series of beds, one above the other, of what appears to be basalt, and these beds with the granites beneath them incline each way from Trout Creek Valley northeast and southwest, forming what appears to be an anticlinal.

As we ascend Trout Creek Pass, we find granites of all textures from very fine compact feldspathic to a coarse aggregate of crystals. There are also many intrusions of trap. All the rocks seem to weather in the same way, by exfoliation, as if it were the desire of nature to round off all sharp points or corners. I think it may be said that Trout Creek Valley is a true anticlinal.

Some time before reaching the top of the pass, we find on the sides of the valley low foot-hills of carboniferous limestone, remnants of a once continuous bed. As we emerge into a little park, just before reaching South Park, we pass through a sort of cañon, with walls of carboniferous limestone on each side, inclining northeast at an angle of eighteen to twenty degrees. This limestone rests directly upon the massive granite, and the bedding of the granite inclines in the same direction and at the same angle. The limestones are from three hundred to four hundred feet in thickness. There is one bed, about thirty feet thick, of rusty quartzose sandstone about the middle of the limestone. The lower beds are very hard, bluish, and cherty; but the upper ones are yellow, purer, and contain imperfect fragments of fossils.

There are here also several examples of the outbursts of basalt, assuming very marked castellated forms.

As we pass into this small park, which is about five or six miles long and two wide, we have on the north side of the road a bed of very thinly laminated black shale, passing up into a great thickness of laminated sandstones, all inclining northeast fifteen degrees, and on the summits of the mountains, four hundred to six hundred feet directly above, are beds of limestone and quartzite inclining in the same direction. The black shales have been prospected for coal. Toward the upper end of this little park is a series of beds, some of them with a reddish tinge, composed of alternate thin beds of shale, sandstone, pudding-stone, and arenaceous limestones, which belong underneath the black shales before mentioned.

It seems to me that these beds are jurassic, or much newer than the carboniferous, but in the upheaval have fallen down below the carboniferous limestones, which have been lifted far up in the ridge beyond. As we ascend the ridge which forms the southwestern rim of the South Park, we meet with what appears to be the same black shales and sandstones on the summit, which we saw some hundreds of feet lower in the small park.

The South Park is completely surrounded with gigantic ranges of mountains, and inside of them the sedimentary rocks, when exposed, seem to dip toward the center of the park. Indeed, I should regard the South Park as one immense quaquaversal.

Around the salt works is a group of laminated sandstones, mostly brown and gray, overlaid by a great thickness of light gray gypsiferous marl with a bed of crystallized gypsum four feet thick. The valley in which the salt springs are located is covered with an efflorescence of what is usually called in this country "alkali." On the east side of the creek which runs past the salt works is a high isolated balsatic butte. About a fourth of a mile east there is a hill composed of the gypsiferous marls, on the surface of which are numerous deposits of calcareous tufa, as if a number of springs had issued from it in former times.

These salt works are quite extensive and costly. The springs are two in number, but the brine is not abundant or strong. Salt has been manufactured here in considerable quantities, and a large portion of Colorado has been supplied with it. These springs are very interesting in a geological point of view, though their origin is somewhat obscure to me, yet I believe they belong to the triassic or saliferous sandstones.

About four miles north of the salt works is a high ridge, inclining northwest twenty degrees, composed of a series of variegated sandstones and shales three hundred to four hundred feet thick. These are, without doubt, the group which I have usually called triassic, or red beds. Still further north we find them inclining southeast, with several thin beds of blue, very hard, cherty limestone, which is characteristic of the red beds. Near Fairplay the brick-red beds are well shown. It seems, therefore, certain that the principal sedimentary rocks which are found in the South Park are triassic.

About ten miles south of Fairplay several thin beds of blue, close, brittle limestone appear, intercalated among the red sandstones, dipping a little east of south, forming a sort of synclinal; that is, the dip is nearly opposite that of the beds near the salt works. These limestones, with the red sandstones, may possibly be of permian age. No fossils could be detected in them. The sandstone is in some cases micaceous, or composed of mica and small crystals of quartz; in others, a fine aggregate of worn pebbles, a sort of fine pudding-stone. These variegated or red beds continue close up to the eruptive ranges for five miles. North and west from Fairplay we come to a high ridge of sandstone with a reddish tinge and slightly calcareous, the dip being north of east, or nearly east, and the ridge forming a marked line running nearly north or south through the middle of the park, from the mountains nearly to the salt works. Just east of this ridge is another ridge of quartzose sandstone or cretaceous. Then comes a very large thickness of the laminated cretaceous clays, covering the country for about fifteen miles. Near McLaughlin's, twelve miles northeast of Fairplay, the lignite tertiary sandstones and clays overlie the cretaceous and jut up against the mountain side. About a mile north of the ranch Mr. McLaughlin has opened a coal mine. He sunk a shaft eighteen or twenty feet through a bed six to ten feet of very impure coal; some portions of it can be used for fuel. The dip of the coal bed is forty-five degrees northeast from the base of the mountains, which are not more than a quarter of a mile distant. Mr. McLaughlin informed me that he had found "oak leaves" in the shale above the coal. These beds occupy the entire north end of the park, and no older rocks are seen between them and the eruptive foot-hills of the mountains. It seems, therefore, that the source of the elevating forces that upheaved these sedimentary formations was in the range of mountains that form the western rim of the park, and, so far as I could ascertain, there are no true ridges of upheaval on the eastern side. Exposures of eruptive rocks are seen everywhere all over the park.

There are several localities where these rocks are thrust up through the cretaceous and tertiary beds, and in the middle and southern portions of the park are quite lofty isolated buttes and mountains of eruptive rocks.

But one of the most conspicuous formations and greatest in extent and importance is the boulder drift. This seems to be mostly confined to the northern and northwestern portions of the park where the principal placer diggings occur. In the valley of the South Platte, especially near Fairplay, there is a prodigious exhibition of the boulder formation. The rocks are well rounded by attrition, and apparently have been swept down from the mountains. Wherever the drift occurs, there are long table lands or terraces, especially in the vicinity of the little streams, and they seem to be planed down with such wonderful smoothness that it must have been done by the combined action of water and ice.

Along the west and north sides of the park are a large number of lofty eruptive peaks, which seem to me old volcanic cones. One of the peaks in the range west of Fairplay seems to have a crater-like summit, the rim broken down on the east side. All around the inside of the remainder of the rim the layers of basalt appear like strata, inclining from the opening in every direction as if the melted material had been poured out and had flowed over the sides in regular strata. There are also tremendous furrows down the sides of others. In the mountains north of the park are huge depressions in these volcanic ranges, the sides of which are quite red, as if they had been in active operation at a comparatively modern period. I am, therefore, inclined to believe that the magnificent range of mountains on the west side of the Arkansas River, extending far northward, is one series of old volcanic cones. As we leave the plains and ascend the mountains at the northeast side of the park, we pass immediately from the older tertiary beds, covered thickly with drift, to the metamorphic rocks mingled with outbursts of eruptive rock. Toward the summit there was a great series of gneissic beds of all varieties and textures. All these mountains east of the park have a gneissic and granitic nucleus. As we descend the valley of a small branch of the North Fork of the South Platte from the Kenosha House, we pass down a monoclinal rift. On the west side is the slope covered with a thick growth of pine and spruce, while on the left side are the projecting edges of the massive red feldspathic granites with two sets of cleavage lines; the vertical with a strike northeast and southwest, and the other inclining at an angle of thirty degrees; the strike, southeast and northwest; while the bedding inclines with the hills. The bedding is so regular and massive that it looks like massive sandstone stratification. The Platte, with all its little branches, flows through these rifts or intervals between the ridges; one side of the stream, a plain gradual slope; the other, extremely abrupt, with the rugged ends of the gneissic or granitic rocks projecting out in a most remarkable manner. After passing along massive granite walls about five miles, we go through four or five miles of singularly banded gneiss, and then massive granite again of every degree of texture, from a fine, close feldspathic rock with no mica, to a coarse aggregate of quartz and feldspar and fine particles of mica. One of the interesting features of these mountains is the fact that all the little streams find their way through these monoclinal valleys. We see also the main axis of the range, composed of massive granite with a distinct bedding, which is sometimes inclined and sometimes horizontal with the banded gneiss inclining from each side. It seems quite clear that each one of these great ranges of mountains is a grand anticlinal with a massive granite axis, with the gneissic granites inclining from each side in the form of ridges, among which the various streams find their way. The trend of these ranges is in the most cases northwest and southeast, or nearly so. Some of the gneissic rocks in the Platte Valley look like laminated sandstone with a regular dip eighteen to thirty degrees. The tops of the highest ranges are, in some cases, covered very thickly with loose fragments of rocks.

Passing down from the junction to Denver we have some of the finest examples of jointage structure in the gneissic rocks that I have ever seen; there are two lines of fracture—one with a direction northeast and southwest, the other northwest and southeast, with the lines of bedding—making a fine study for the geologist. Some of the beds are thus broken into nearly square blocks, and others in diamond-shaped masses.

On reaching the base of the mountains the usual ridges of sedimentary rocks are passed over—red beds, jurassic, cretaceous, and tertiary. The tertiary beds commence within a mile of the foot of the mountains, soon becoming horizontal in their position, and before reaching Denver they are scarcely seen on account of the superficial deposit of drift and alluvial which covers them.

CHAPTER X.

TRIP TO MIDDLE PARK.

Our route to the Middle Park was through the Berthoud Pass, from the valley of Clear Creek. The range of mountains in which the pass is located is composed of gneissic rocks—as are all the ranges in the mining districts. The ascent was very steep on the south side, up to the region of perpetual snow; but the descent on the north side is quite gradual.

Great quantities of loose materials from the basis rocks are scattered thickly over the summits, of every variety of the metamorphic class. Most of the peaks are well rounded, and covered with soil and vegetation. Grass and flowers grow far up above the limits of arborescent vegetation. As we ascend, the pines, spruces, and cedars dwindle down in size until they become recumbent and trail on the ground. Some of the highest peaks are very sharp and covered with loose rocks, as if only the usual atmospheric influences had ever affected them. Their sides are often massive escarpments of rocks down which an infinite quantity of fragments have fallen, making a vast amount of debris at the base. Of course their rocky sides are entirely free from vegetation, and the oxide of iron gives them a rusty reddish appearance. One mountain at the head of Clear Creek is called Red Mountain, from the fact that the rocks have a bright red color in the distance. The evidences of the outpouring of igneous rocks in this mountain are very marked; indeed, it may be called an eruptive range.

From the summit of Berthoud's Pass, at a height of eleven thousand eight hundred and sixteen feet, we can look northward along the line of the main range, which gradually flexes around to the northwest, while the little streams seem to flow through the rifts. The general appearance of the western slope of this great range would indicate that it is a huge anticlinal composed of a series of ranges on each side of a common axis, and then smaller ranges ascend like steps to the central axis. The western side of this ridge slopes gently, while the eastern side projects over abruptly. This main range also forms a narrow dividing line, or "water-divide," between the waters of the Atlantic and Pacific. I stood where the waters of each side were only a few feet apart, and felt a real joy in passing down the western slope of the mountain by the side of a pure crystal stream whose waters were hastening on to the great Pacific.

All down the western slope is a great thickness of superficial material, loose sand, decomposing feldspar, with partially worn rocks of all sizes. This is due quite evidently to local influences, ice and water wearing down the sides of the mountains and depositing the material adhering to the masses of ice along the slopes.

The springs of water are very numerous, and the water seems to collect in the thick grass and moss-covered earth, forming large bogs. It is also interesting to watch the growth of a stream from its source, receiving in its way the waters of myriads of springs, until it becomes a river too formidable to ford easily. The little stream which rises in the pass we followed to the Park, where it is fifty yards wide, and contains an abundance of fine trout.

The Middle Park is really made up of a number of smaller parks, which are somewhat independent of each other. Each one may present different geological formations. The little park on the south side, which we first enter, is a very beautiful one. The grass is luxuriant, and the timber excellent. None of the older sedimentary rocks were seen along the flanks of the mountains, but a recent tertiary deposit seemed to cover the country. On the east side of Fraser Creek there is a long, high ridge, which is cut by the stream in several places, formed of the white and yellow sands and marls which mark the pliocene tertiary on the east side of the mountain. I have no doubt that it is a formation of the same kind as that of the Arkansas marls, and cotemporaneous with it.

Along this creek there are some massive walls of this formation, mostly yellow marls, but some layers of sandstone. This ridge extends from the mountains far northward, and is about two miles wide; and between it and the immediate base of the mountains is situated a beautiful valley of considerable width.

The Middle Park is apparently a quaquaversal, surrounded by the lofty snowy ranges; and the lower ranges descending like steps to the valley which constitutes the true park. The park does not appear to be more than from ten to twenty miles wide from east to west, and from fifty to sixty long from north to south. In this park also the ranges of mountains so surround it that the slopes seem to form a sort of quaquaversal inclining toward a common center.

Viewed from Middle Park, Long's Peak, and the range immediately connected with, has a rugged, saw-like edge, as if composed of eruptive rocks, and ridge after ridge inclines from it in regular order.

About ten miles north of our camp, in the first park, we come to low ridges of massive red feldspathic granite, and parallel with these granite ridges are a series of sedimentary beds, commencing with the brick-red beds. The strike is nearly north and south, and the dip west. These ridges are all so grassed over that the true nature of the underlying rocks is not easily determined. Then comes ridge after ridge until all the beds—jurassic and cretaceous—are shown.

On this stream we have a fine system of terraces. On the north side are three distinct terraces above the bottom, and the lowest one has a bed of cretaceous sandstone, nearly horizontal, cropping out at its base. This is a low one, not more than fifteen feet high; the next one is fifty feet high, and the third, which descends from the high hills, is two hundred feet. A little west of south, at the junction of Grand River with Fraser Creek, five high peaks are visible, which form in that direction a part of the main range. All around us, in every direction, we could see the snowy peaks, and the beds which form the ridges of upheaval inclining in every direction.

To the south of the park the older sedimentary rocks dip north in lofty ridges, at least two thousand feet high, presenting high escarpments when split by streams, and reaching almost the highest margin of the mountains.

About ten miles above the hot springs, Grand River flows through an enormous gorge cut through a high ridge of basalt, which seems to be an intrusive bed, for above and below, the sedimentary rocks are well shown, but partially changed. Underneath are the cretaceous shales of Nos. 4 and 5, and above are the lignite tertiary beds. These beds all dip west twenty-three degrees.

These eruptive rocks are very rough, as if they had been poured out without much pressure. Much of it is a very coarse conglomerate, the inclosed masses appearing to be the same kind as the paste; that is, originally, of igneous origin. Some of the inclosed rocks are very compact, close, and all were, more or less worn before being inclosed. This rock is a true dolorite. I did not see any inclosed masses that I could call unchanged. This basalt extends a great distance, continuing a nearly uniform thickness, and inclining in the same direction with the cretaceous beds below and the tertiary beds above.

On both sides of Grand River, but especially on the east and northeast sides, extending up nearly to the foot of Long's Peak, are quite large exposures of the recent tertiary beds. They are nearly horizontal, and have much the appearance in color of the Fort Bridger beds, of which Church Buttes is an example. These beds are composed, for the most part, of fine sand and marl, but there are a few small rounded boulders scattered through it. Below the gorge, on the north side of Grand River, these outflows of basalt have formed some well-defined mesas; at least three beds ascending like steps from the river. Below the gorge the river flows through what seems to be a rift of basalt, that is, on the north side. The basalt lies in horizontal beds, but on the south side is the sloping side of a basaltic ridge. The dip is nearly northwest, though the trend of this basaltic ridge is by no means regular. One portion of it has a strike northwest and southeast, and another north and south. The tertiary rocks reach a great thickness, and are elevated high up on the top of the basaltic ridge, eight hundred to one thousand feet above the river. They are mostly formed of fine sandstone and pudding-stone. These fine sandstones contain some well-marked impressions of deciduous leaves, among which are good specimens of *Platanus haydeni*. On the north side of Grand River, in some localities, the tertiary beds are elevated so high, on many of the eruptive mountains, that they are covered with perpetual snow. These eruptive beds are certainly among the most remarkable examples of the overflow of igneous matter that I have ever seen in the West.

At one locality I saw a remarkable intrusive layer between the red or variegated beds which are supposed to be triassic and the jurassic. It is a very compact, heavy syenite, and forms a ridge of upheaval, and dips in the same direction and at the same angle with the unchanged beds above and below.

About four miles below the first basaltic cañon on Grand River, apparently the same ridge comes close to the river again. On the north side there is a high basaltic uplift, which shows well marked lines of stratification, as if the melted material had been poured out in thin regular sheets or layers. The dip is about north. In many places the entire mass is made up of a coarse conglomerate, and has the peculiar steel color which seems to characterize modern eruptive rocks. The dip of this basaltic ridge, at this point, is thirty-six degrees. On the opposite side of the river there is an isolated portion cut off from the main ridge, with a dip about south or southeast twenty-four degrees.

Continuing our way west down Grand River we pass over a series of upturned ridges of sedimentary rocks, inclining in the same direction with the basaltic ridge trending parallel with it, composed of cretaceous and older tertiary beds. Looking eastward from the Grand Cañon, below the hot springs, this remarkable basaltic ridge seems to form a semi-circle with a general dip about north.

Immediately below the hot springs the Grand Cañon commences, and the river cuts its way through an upheaved ridge of massive feldspathic granite for three miles between walls from one thousand to one thousand five hundred feet high. The south side is somewhat sloping and covered thickly with pines, while the north side is extremely rugged, the immense projecting masses of granite forbidding any vegetation to gain a foothold. It would seem that the river had worn its way through a sort of rift in the granite, but at the upper end it has cut through the uplifted sedimentary ridges nearly at right angles. In some places the north side is gashed out in a wonderfully picturesque manner, so that isolated columns and peaks are left standing, while all the intermediate portions have been worn away. This granite ridge will average perhaps five miles in width, and extends an unknown distance across the park northeast and southwest, and it is from the southeast side that the ridges of upheaval above described incline.

The granite ridge seems to form a sort of abrupt anticlinal. On the southeast side the rocks are all bare or covered with a superficial deposit of recent tertiary marls. None of the older unchanged rocks are seen on this side, but the modern sands and sandstones are exposed in a horizontal position in the channel of the river.

The hot springs are located on the right bank of Grand River, at the juncture of the sedimentary rocks with the granites. Just east of the springs is a high hill, Mount Bross, one thousand to one thousand two hundred feet above Grand River, which seems to be composed mostly of the older tertiary strata, alternate yellow and gray sandstones, and laminated arenaceous shaly clays. The whole is so grassed over that it is difficult to take a section. The beds incline east of north at a small angle. I regard the beds as of the age of the coal formations of the West, older tertiary. I found excellent impressions of deciduous leaves, among which are those of the genus *Magnolia*. Just opposite the springs, the left bank of the river shows a perfect section of all the layers from the cretaceous to the jurassic. The bank is not more than ten feet thick above the water, and yet it shows that the river itself rolls over the upturned edges of all these beds.

The section in descending order is as follows:

1. Tertiary strata forming the greater part of the hill known as Mount Bross.

2. Gray laminated sandstones passing down into arenaceous clays with *Baculites ovatus*, &c.

3. Black clays of No. 4. These are of great thickness and every variety of texture. As shown in a cut bank of the river it is yellow arenaceous clay with layers of sandstone, in which the impressions of deciduous leaves were observed. These layers project up, a distance along the bank, of seventy paces.

4. Dark plastic clay with cone in cone, seams of impure clay, iron ore. Then comes an interval in which no layers could be seen, sufficient to include No. 3—two hundred and fifty paces.

5. Dark steel-black laminated slate, with numerous fish scales; dip, twenty-seven degrees. This slate passes down into alternate layers of rusty sandstone and shaly clay.

In the upper bed of sandstone and shaly clay are obscure vegetable impressions, leaves, stems, nuts, &c., evidently deciduous. In the upper bed of sandstones are two or three thin seams of carbonaceous shale, and the intervening layers of sandstone are almost made up of bits of vegetable matter. Toward the lower, it becomes a hard mud rock passing down into rusty yellow sandstone with all sorts of mud markings. Then comes a bed of bluish plastic clay with sulphur and oxide of iron; dip, thirty-three degrees. Then rusty fine-grained gray sandstone passing down into a very close massive pudding-stone, composed of very smooth nicely-rounded pebbles, cemented with silica. This stone would be most excellent for building material and is susceptible of a very fine polish. A fracture passes directly through the pebbles, the paste being harder, if anything, than the inclosed pebbles; dip, thirty-one degrees. This is a very thick bed and is a portion of No. 1, cretaceous, or a sort of transition bed between the cretaceous and the jurassic.

The red and variegated beds lie fairly upon the gneissic granites, and although they are shown very obscurely here, yet I think they must exist, inasmuch as they are so well revealed not more than fifteen miles east of this point, so that I have no doubt they are lost beneath the mass of superincumbent material. I think the light-colored clays lying underneath the bed of chalky clay, are jurassic. There is a bed of fine gritty clay underneath the pudding-stone which would make excellent hones.

In the intercalated sandstones above the pudding-stones are plants just like those observed in No. 1 at Sioux City, on the Missouri River, and the composition of the strata is the same; there is a *Salix*, a coniferous plant, the cones of a pine, &c.

I have given this detailed description of the cretaceous rocks to show the exceeding variableness of their texture, and also to call the attention of scientific men, who may hereafter visit this interesting locality, which will soon become celebrated, to a section of the rock through which the waters of the spring must pass in reaching the surface. Now in whatever rocks these springs may originate, the water must pass a long distance through the almost vertical strata of the cretaceous period, in the sediments of which are found in other localities nearly all and perhaps all the mineral constituents found in these springs. The deposits around these springs are very extensive. No analysis has as yet been made, but large masses of gypsum and native sulphur can be taken out at any time from the sides of the large basin-like depression into which the water flows. They are properly "Hot Sulphur Springs," varying in temperature from eighty to one hundred and twelve degrees.

About fifteen miles west of the springs is the valley of the Troublesome Creek, a small branch of the Grand River, flowing from the basaltic mountains on the northern side of the park.

I visited this region under the guidance of Mr. Sumner, an old resident of the park. The surface of the country along our road was strewn with eruptive rocks. We saw several localities where the basaltic rocks protruded, and one place in Corral Creek, about eight miles west of Grand River, where the little stream has cut a deep channel through the red granites. The older tertiary beds appear from time to time.

Troublesome Cañon, at the head of the creek bearing this name, is entirely basaltic, and the rugged walls not only of the main stream but also of the little branches, form a most picturesque view.

Below the cañon, the valley of Troublesome Creek, and also that of Grand River near the junction, is occupied by belts of modern tertiary sands and marls like those observed at the entrance to the park, by Berthoud's Pass. Where the little stream cuts the terraces, horizontal strata of whitish and flesh-colored sands and marls are exposed. I looked in vain for fossils and found only specimens of silicified wood. There are cold sulphur springs in this valley. All through the park, the benches or terraces are conspicuous in the vicinity of streams, as at the base of mountain ranges. In the park through which Frazer's Creek flows, these benches or terraces are most beautifully carved out from the modern marls.

I regret that my visit to the Middle Park was so short that I could not explore the entire area with care, for few districts in the West can afford more material of geological interest, and an entire season could be spent studying its geology and geography with great profit.

The agricultural resources of the Middle Park are as yet unknown. No attempt has been made to cultivate any portion of it. Grass and grazing are excellent and the soil good, and if the climate will permit, all kinds of garden vegetables could be raised in abundance, and some varieties of the cereals. Timber is abundant both for lumber and fuel.

In summing up the geology of the Middle Park, we find that all the sedimentary rocks known in this country are found there. I did not see any beds that I could define as carboniferous, but the triassic, jurassic, cretaceous, and tertiary are well developed. I have no doubt as to the existence of true carboniferous limestones in the Middle Park.

The tertiary deposits of this region may be divided into two groups, viz, the lignite or older tertiary, and the modern pliocene marls and sands which seem common to the parks and mountain valleys. The former conform perfectly to the older beds, while the latter seldom incline more than three to five degrees, and do not conform to the older rocks. The marl group is undoubtedly contemporaneous with the Arkansas and Sante Fé marls.

The geological structure of the Middle Park is more varied, compli cated, and instructive than that of any other of the parks.

CHAPTER XI.

THE GOLD AND SILVER MINES OF COLORADO.

I will confine my remarks mostly to the geological features connected with these mines, inasmuch as Mr. Frazer, in his report appended, has fully treated this subject.

The gold and silver lodes of this Territory, so far as they are observed, are entirely composed of the gneissic and granite rocks, possibly rocks of the age of the Laurentian series of Canada. At any rate, all the gold-bearing rocks about Central City are most distinctly gneissic, while those containing silver at Georgetown are both gneissic and granitic. The mountains in which the Baker, Brown, Coin, Terrible, and some other rich lodes are located, is composed mostly of gneissic and reddish feldspathic granite, while the Leavenworth and McClellan Mountains, equally rich in silver, are composed of banded gneiss, with the lines of bedding or stratification very distinct.

There is an important question that suggests itself to one attempting to study the mines of Colorado, and that is the cause of the wonderful parallelism of the lodes, the greater portion of them taking one general direction or strike, northeast and southwest. We must at once regard the cause as deep-seated and general, for we find that most of the veins or lodes are true fissures and do not diminish in richness as they are sunk deeper into the earth. All these lodes have more or less clearly defined walls, and some of them are quite remarkable for their smoothness and regularity. We assume the position that the filling up of all these lodes or veins with mineral matter was an event subsequent to any change that may have occurred in the country rock. Now, if we look carefully at all the azoic rocks in this region we shall find more or less distinctly defined, depending upon the structure of the rock itself, two planes of cleavage, one of them with a strike northeast and southwest, and the other southeast and northwest. Beside these two sets of cleavage planes there are in most cases distinct lines of bedding. The question arises, what relation do these veins hold to these lines of cleavage? Is it not possible that they occupy these cleavage openings as lines of greatest weakness?

I have taken the direction of these two sets of cleavage planes many times with a compass, over a large area, and very seldom do they diverge to any great extent from these two directions, northeast and southwest or southeast and northwest. In some instances the northwest and southeast plane would flex around so as to strike north and south, and the other one so as to trend east and west, but this is quite seldom, and never occurs unless there has been some marked disturbance of the rocks. There are, however, a few lodes which are called "east and west lodes," and some, "north and south." A few have a strike northwest and southeast, but are generally very narrow and break off from the northeast and southwest lodes, are very rich for a time and then "pinch" out. It would seem therefore quite possible that the northeast and southwest veins took the lines of cleavage in that direction as lines of greatest weakness, and that the northwest and southeast lines cross the other set, and that a portion of the mineral material might accumulate in that cleavage fissure. I merely throw out this as a hint at this time, which I wish to follow out in my future studies. I am inclined to believe that the problem of the history of the Rocky Mountain ranges is closely connected with these two great sets of cleavage lines. As I have before stated, my own observations point to the conclusion that the general strike of the metamorphic ranges of mountains is northwest and southeast, and that the eruptive trend northeast and southwest. The dikes that sometimes extend long distances across the plains, in all cases trend northeast and southwest, or occasionally east and west. The purely eruptive ranges of the northern portion of the San Luis Valley seem to be composed of a series of minor ranges "*en èchelon*" with a trend northeast and southwest. But as soon as this range joins on to a range with a metamorphic or granitic nucleus, the trend changes around to northwest and southeast. Many of the ranges have a nucleus of metamorphic rocks though the central and highest portions may be composed of eruptive peaks and ridges. In this case the igneous material is thrust up in lines of the same direction as the trend. It becomes therefore evident that all the operations of the eruptive forces were an event subsequent to the elevation of the metamorphic nucleus. This is shown in hundreds of instances in Southern Colorado and New Mexico, where the eruptive material is oftentimes forced out over the metamorphic rocks, concealing them over large areas.

All over the mining districts are well-marked anticlinal, synclinal, and what I have called monoclinal valleys. Nearly all the little streams flow a portion or all their way through these monoclinal valleys or rifts. In most cases the streams pass along these rifts from source to mouth, but occasionally burst through the upheaved ridges at right angles, and resuming its course again in some monoclinal opening. There are a few instances of these streams flowing along anticlinal valleys, and by any one these remarks will be at once understood by studying the myriad little branches of Clear Creek or South Platte, which flow for long distances through the mining districts.

In these valleys are oftentimes accumulated immense deposits of modern drift. Sometimes there are proofs that these valleys have been gorged for a time, and a bed of very coarse gravel and boulders will accumulate, hundreds of feet in thickness. Near Georgetown there is a fine example of this modern drift action.

It would seem that the valley of that branch of Clear Creek, in which the Brown and Terrible silver lodes are located, was gorged at one time, perhaps, with masses of ice, and the fine sand and coarse materials accumulated against the gorge, and at a subsequent period the creek wore a new channel through this material. The upper side of this drift deposit is fine sand, but the materials grow coarser as we descend, until, at the lower side, there are immense irregular or partially worn masses of granite. On the sides of the valley the rocks are often much smoothed and grooved as if by floating masses of ice. We assume the position, of which there is most ample evidence all over the Rocky Mountain region, that at a comparatively modern geological period the temperature was very much lower than at present, admitting of the accumulation of vast bodies of ice on the summits of the mountains. The valley of the South Platte, as that stream flows through the range east of the South Park, show, not only these accumulations of very coarse boulder drift, but when this drift is stripped off, the underlying rocks are found smoothed, and, in some instances, scratched, as if by floating icebergs.

In regard to the character of the gold and silver mines of Colorado, much information of practical value has been secured, but my limited time will not permit me to present it in detail in this preliminary report.

It will be more fully elaborated during the coming winter. I would simply remark that my observations indicate to me that the silver mines of Georgetown are very rich and practically inexhaustible, and that, under the present system of working them, they are becoming daily more and more important. The amount of labor that is continually expended in opening mines and driving tunnels is immense, and their future importance as a source of wealth to the country greatly increased. The same remarks will apply to the gold mines of Gilpin County. There are some remarkably rich lodes which have yielded the enterprising miners untold wealth, and some that will continue to do so. In the majority of cases, where proper management and economy have been employed, the mines have been a great source of profit to the miner. It is not necessary to enter into the causes of the wonderful failures and swindling operations which have brought Colorado into such disrepute in the past. It is sufficient for me to state my belief that the mining districts of Colorado will yet be regarded as among the richest the world has ever known.

CHAPTER XII.

REVIEW OF LEADING GROUPS, ETC.

This final chapter to my report, which I have added here, will contain a brief review of the leading groups of strata noticed in this and my previous reports, as well as a few additional observations and chemical analyses. The details of my labors will be presented in my final reports at some future period.

I have already alluded to my belief that this western country during the tertiary period was covered to a greater or less extent with a chain of brackish or fresh-water lakes; that the tertiary period began its existence with brackish water deposits, which gradually became fresh water, and thus continued up to the present time. It is hardly possible to synchronize all these groups of strata with our present knowlege; but in order that our efforts in that direction may be facilitated, I have thought it best to give them specific names, which may be regarded provisional for the present. Each one of these groups will doubtless afford a flora and fauna to a certain extent peculiar to itself, and a greater importance will be attached to it when grouped around some specific names.

Proceeding southward from Cheyenne we pass over the coal formations of the tertiary period, which have already been called, on the Upper Missouri, the Fort Union group. This group I regard as marking the dawn of the tertiary age in the West, and as covering a far more extended area than any other group of this epoch. It is continuous southward from the Missouri Valley to Colorado, interrupted only by a belt of White River beds about two hundred miles wide. I think these beds also extend far northward into the British possessions, probably nearly or quite to the Arctic Sea.

About forty miles south of Denver we have a high divide, or ridge, which forms a sort of water-shed between the Platte and Arkansas Rivers. This is composed of a group of strata, mostly sandstones and sands jutting up against the mountains in a slightly disturbed position and not conforming to the older rocks. These beds are undoubtedly middle tertiary, and I have called them the Monument Creek group.

I do not think that such terms as eocene, miocene, pliocene, &c., are at all applicable to the tertiary deposits of the West, and I therefore designate them as lower, middle, and upper tertiary. I regard all the coal beds of the West as lower tertiary. It is true that some of these beds of lignite, or impure coal, or carbonaceous clay, are found in groups of strata which should be classed as middle tertiary; but these do not seem in any case to be of any economical importance.

Near Hard Scrabble Creek, a small branch running into the Arkansas River just below Cañon City, there is a small area, about eight miles square, occupied by coal strata, for which I propose the provisional name of Cañon City group. I have but little doubt that careful study will show that it is a fragment of the great lignite group of the North. The next group comprises the coal beds of the Raton Hills, which I suspect is also a portion of the great lignite group, and will eventually be found to be synchronous with it. I have called it the Raton Hills group.

The next group of coal strata occurs in Placer Mountains, New Mexico, about thirty miles south of Santa Fé. The lithological character of the beds or rocks are very similar to those of the lignite group further north, but the evidence in regard to their age, or parallelism with the lignite group, is not so clear. While I regard the true coal beds of the West as lower tertiary, yet these Placer Mountain beds present the appearance of greater antiquity than the coal beds further north. Still, the numerous varieties of deciduous leaves which I have obtained from rocks just overlying the coal beds indicate that they are lower tertiary; and with this belief I have named them the Placer Mountain group.

Overlying the Placer Mountain beds, in the valley of Gallisteo Creek, is a vast thickness of exceedingly variegated sands, sandstones, and calcareous sandstones, characterized mostly by containing an abundance of silicified wood; but no other fossils have, as yet, been discovered. I have given this series of beds the name of Gallisteo sands, and they are doubtless middle tertiary.

In the valley of the Rio Grande, at least from Albuquerque to the north end of San Luis Valley, a series of marly sands of a light color prevail to a greater or less extent. They exhibit their greatest thickness north of Santa Fé. To this group I have given the name of Santa Fé marls; and they are doubtless of the age of upper tertiary, and synchronous with the upper beds of the White River group as seen along the North and South Forks of the Platte and near Cheyenne.

In the valley of the Arkansas, north of the Poncha Pass, is a fine development of the light-colored marls, doubtless of the same age with the Santa Fé marls, which I have designated by the name of the Arkansas marls. I have as yet obtained no well-defined fossils from either the Santa Fé or Arkansas marls; yet bones of some large animal, probably mastodon or elephant, have been found in them. I have no doubt that more careful explorations will show that a fauna and flora of greater or less extent will characterize all these groups.

Along the Union Pacific railroad we find in the Laramie Plains a most extensive exhibition of the great lignite group. The first coal beds of great economical value occur near Carbon and at Separation. From Creston to Bitter Creek there are a series of purely fresh-water beds, with some beds of impure lignite, with vast quantities of fossils belonging to the genera *Unio*, *Melania*, *Vivipara*, *Helix*, &c. This group I regard as middle tertiary, and the strata are very nearly horizontal. I have regarded these beds as separated from the lower tertiary or true lignite group, and have designated them by the name of the Washakee group. A little east of Rock Spring station a new group commences, composed of thinly laminated chalky shales, which I have called the Green River shales, because they are best displayed along Green River. They are evidently of purely fresh-water origin, and of middle tertiary age. The layers are nearly horizontal, and, as shown in the valley of Green River, present a peculiarly banded appearance. When carefully studied these shales will form one of the most interesting groups in the West. The flora is already very extensive, and the fauna consists of *Melanias*, *Corbulas*, and vast quanties of fresh-water fishes, preserved in much the same way as those in the Solenhofen slates of Germany. There are also numerous insects and other small undetermined fossils in the asphaltic slates. One of the marked features of this group is the great amount of combustible or petroleum shales, some portions of which burn with great readiness, and have been used for fuel in stoves.

The next group commences not far west of Bryan, and is doubtless a prolongation upward of the Green River shales, and may be regarded as of upper tertiary age.

The sediments are composed of more or less fine sands and sandstones, mostly indurated, sometimes forming compact beds, but usually weathering into those castellated and dome-like forms which have given such celebrity to the "Bad Lands" of White River. Church Buttes, near Fort Bridger, is an example of this group, and shows the style of weathering to which I refer. I have called this series of beds the Bridger group, from the fact that it is best developed in this region. It has already yielded remarkably fine species of *Unio*, *Melania*, *Planorbis*, *Vivipara*, *Helix*, &c., with a great variety of turtles and mammalian remains. There are indications that when this group is thoroughly explored it will prove to be second only to the "Bad Lands" of Dakota in the richness and extent of the vertebrate remains.

Immediately west of Fort Bridger commences one of the most remarkable and extensive groups of tertiary beds seen in the West. They are wonderfully variegated, some shade of red predominating. This group, to which I have given the name of Wasatch group, is composed of variegated sands and clays. Very little calcareous matter is found in these beds.

In Echo and Weber Cañons are wonderful displays of conglomerates, fifteen hundred to two thousand feet in thickness. Although this group occupies a vast area, and attains a thickness of three to five thousand feet, yet I have never known any remains of animals to be found in it. I regard it, however, as of middle tertiary age.

After passing Rock Springs station, Union Pacific railroad, the next exposures of coal are at Bear River City, and at Evanston, and also at Coalville, near the entrance of Echo Creek into Weber River. The coal beds at Evanston are the finest known in the West, and reach a thickness of twenty-six feet at one locality. These coal beds seem to be separated from those at Separation and Carbon, and to present some features different from those in any other portion of the West. I am in doubt as to their precise position, but I am inclined to regard them as of lower tertiary age, possibly on a parallel with the oldest beds of the great lignite group in other localities. On Bear River we find several species of *Ostrea*, both above and below the coal, and in a cut just west of Bear River City is found the greatest profusion of molluscous life that I have ever seen in any of the tertiary beds of the West. There seems here to be a mingling of fresh and brackish water fossils. At Evanston, impressions of deciduous leaves are abundant in beds above the coal. No portion of the fauna seems to be identical with anything found in other places. The flora seems also to be distinct, although some of the forms may be identical with species elsewhere. I have named the group of coal strata which is exposed from beneath the middle tertiary beds by upheaval at Bear River City, Evanston, and Coalville, the Bear River group.

In the valley of Weber River, from Morgan City to Devil's Gate, there is a thickness of one thousand to twelve hundred feet of sands, sandstones, and marls, of a light color for the most part, which I regard as of upper tertiary age. These newer beds must have not only occupied this expansion of the Weber Valley, but also all of Salt Lake Valley, for remnants of it are seen all along the margins of the mountains inclosing Salt Lake Valley. I have obtained one species of helix near Salt Lake City from this group, which very much resembles a species obtained from the Wind River deposits, near the source of Wind River. I found this series of beds so widely extended and so largely developed in Weber Valley and Salt Lake Valley, that I regard it as worthy of a distinct name, and in consequence have called it the Salt Lake group.

Some years ago, in a paper published in the proceedings of the Academy of Natural Sciences, at Philadelphia, Mr. Meek and the writer proposed names for certain groups of tertiary strata, which might be added to the list already given:

First. The Fort Union or great lignite group, which occupies the whole country around Fort Union near the mouth of the Yellowstone, extending north into the British possessions to unknown distances, also southward on the Missouri River to Fort Clark. It also extends along the eastern flanks of the mountains, probably to Denver, Colorado, and perhaps further.

Second. The Wind River deposits are limited, so far as we now know, to the Wind River Valley. The sediments are composed of indurated sands and clays, with a few layers of sandstones and some calcareous concretions; and the prevailing color is very light gray, sometimes brown with reddish bands. The fossils thus far found are fragments of *Trionyx*, *Testudo*, *Helix*, *Vivipara*, petrified wood, &c., doubtless of middle tertiary age.

Third. The White River group, best shown on White River, Dakota, but covering a very extended area—at least one hundred and fifty thousand square miles. The sediments are composed of white and light drab indurated sands, clays, and marls, with some beds of sandstones and limestones; is purely fresh water, and remarkable as one of the most wonderful deposits of extinct mammalia on the globe—middle tertiary.

Fourth. The Loup River beds, which certainly form a most singular and remarkable group. They are composed for the most part of fine, loose gray or brown sands, with some layers of limestone containing a distinct and most remarkable fauna, composed of wolves, foxes, tigers, hyenas, camels, horses, mastodons, elephants, &c. There are also numerous fresh-water and land shells, perhaps of recent species, upper tertiary. To these groups might be added the Judith River beds, a small basin on the Missouri River, near the foot of the mountains, about fifteen to twenty miles in width and forty miles in length. This group is probably of lower tertiary age, but I think it was always separated from the great lignite group. In my final report I hope to be able to illustrate each one of these groups by the organic remains peculiar to it, and, if possible, show the relations of each one to the other and to all. Further explorations of the Territories will reveal many more of these lake basins, for I am now convinced that all over the great area west of the Mississippi to the Pacific coast the evidence of the existence of these lakes will be more or less clear.

DEPOSITS OF COAL AND IRON ORE.

One of the most important problems to be solved in the West is the utilization of the vast quantities of iron ore which are scattered all over the country in a multiplicity of forms. The brown iron ores accompany the coal beds everywhere, and some good deposits are found in the cretaceous formations. At the source of the Chugwater are immense deposits of magnetic iron ore in the metamorphic rocks, which are probably of Laurentian age, while at Rawlings's Springs are most valuable beds of the red oxide of iron, in rocks which I suppose to be of triassic age. The latter are evidently local, but the amount of iron ore is considerable.

The following extract is taken from the excellent report of Dr. John L. Leconte to the Pacific Railroad Company:

"Deposits of iron ore fit for working are found in the sandstones of the Vermejo, as described on page 24 of the first part of the report. Veins of specular, titaniferous, and magnetic ore, occur in the metamorphic rocks of the mountains; those near Vegas are mentioned on page 29. Large quantities of magnetic iron are found near the Ortiz mine, and beds of an argillaceous variety occur near the anthracite of the Placer Mountain, as mentioned on page 39.

"Should the coal be capable of use for smelting iron, the localities of the latter will be found ample for all possible demands.

"I have received from Messrs. Williams and Moss the following results of the examination of some iron ores collected on the journey:

1. Magnetic iron ore, Las Vegas, metallic iron........ 20.43 per cent.
2. Magnetic iron ore, Placer Mountain, metallic iron.. 65.27 per cent.
3. Carboniferous iron ore, Vermejo Cañon............. 21.91 per cent.
4. Carbonate of iron, near anthracite of Placer Mountain 36.49 per cent.

I take the liberty of introducing in this connection the following extracts from an article written by me and published in Silliman's Journal, March, 1868. This paper has been very extensively copied, and even now I find it necessary to make but few changes:

Mines have been opened on Coal Creek, three miles south of Marshall's mines, but they have been abandoned for the present. Another has been opened about twenty miles south of Cheyenne City, on Pole Creek. The drift began with an outcropping of about four feet eight inches in thickness, inclination twelve degrees east. The lignite grows better in quality as it is wrought further into the earth, and the bed, by following the dip two hundred feet, is found to be five feet four inches thick, and the lignite is sold readily at Cheyenne City for twenty-five dollars per ton. The beds are so concealed by a superficial drift deposit, that it is difficult to obtain a clearly connected section of the rocks. A section across the inclined edges of the beds eastward from the mountains is as follows:

7. Drab clay passing up into areno-calcareous grits composed of an aggregation of oyster shells, *Ostrea subtrigonalis*.
6. Lignite—5 to 6 feet.
5. Drab clay—4 to 6 feet.
4. Reddish, rusty sandstone in thin laminæ—20 feet.
3. Drab arenaceous clay, indurated—25 feet.
2. Massive sandstone—50 feet.
1. No. 5 cretaceous, apparently passing up into a yellowish sandstone.

The summit of the hills near this bed of lignite is covered with loose oyster shells, and there must have been a thickness of four feet or more, almost entirely composed of them. The species seems to be identical with the one found in a similar geological position in the lower lignite beds of the Upper Missouri near Fort Clark, and at the mouth of the Judith River, and doubtless was an inhabitant of the brackish waters which must have existed about the dawn of the tertiary period in the West. No other shells were found in connection with these in Colorado, but on the Upper Missouri well-known fresh-water types exist in close proximity, showing that if it proves anything, it rather affirms the eocene age of these lower lignite beds. These lignite beds are exposed in many localities all along the eastern base of the mountains, and from the best information I can secure, I have estimated the area occupied by them north of the Arkansas River at five thousand square miles. According to the explorations of Dr. John L. LeConte during the past season, which are of great interest, these same lignite formations extend far southward into New Mexico on both sides of the Rocky Mountains. Specimens of lignite brought from the Raton Mountains by Dr. LeConte, resemble very closely in appearance and color the anthracites of Pennsylvania. It is probable that no true coal will ever be found west of longitude ninety-six degrees, and it becomes therefore a most important question to ascertain the real value of these vast deposits of lignite for fuel and other economical purposes. Can these lignites be employed for generating steam and smelting ores? In regard to the lignites in the Laramie Plains, I have as yet seen no analysis, but specimens are now in the hands of Dr. Torrey, of New York, for that purpose; specimens from Marshall's mine on South Boulder Creek were submitted to Dr. Torrey by the Union Pacific Railroad Company for examination, with the following result:

Water in a state of combination, or its elements	12.00
Volatile matter expelled at a red heat, forming inflammable gases and vapors	26.00
Fixed carbon	59.20
Ash of a reddish color, sometimes gray	2.80
	100.00

A specimen from Coal Creek, three miles south, yielded similar results:

Water in a state of combination, or probably its elements, as in dry wood	20.00
Volatile matter expelled at a red heat in the form of inflammable gases and vapors	19.30
Fixed carbon	58.70
Ash, consisting chiefly of oxide of iron, alumina, and a little silica	2.00
	100.00

The percentage of carbon is shown to be in one case 59.20, and in the other 58.70, which shows at a glance the superiority of the western lignites over those found in any other portion of the world. Anthracite is regarded as so much superior as a fuel, on account of the large per cent. of carbon, and also the small amount of hydrogen and oxygen. The bituminous coals contain a large percentage of hydrogen and oxygen, but not enough water and ash to prevent them from being made useful, but the calorific power of lignite is very much diminished by the quantity of water contained in it, from the fact that so valuable a portion of the fuel must be used in converting that water into steam.

The day of my visit to the Marshall coal mines, on South Boulder Creek, seventy-three tons of lignite were taken out and sold at the rate of four dollars a ton at the mine, and from twelve to sixteen dollars at Denver. This lignite is somewhat brittle, but has nearly the hardness of ordinary anthracite, which it very much resembles at a distance.

In some portions there is a considerable quantity of resin. I spent two evenings at Mr. Marshall's house, burning this fuel in a furnace, and it seemed to me that it would prove to be superior to ordinary western bituminous coals, and rank next to anthracite for domestic purposes. Being non-bituminous, it will require a draught to burn well. It is as neat as anthracite, leaving no stain on the fingers. It produces no offensive gas or odor, and is thus superior in a sanitary point of view, and when brought into general use, it will be a great favorite for culinary purposes. It contains no destructive elements, leaves very little ash, no clinkers, and produces no more erosive effects on stoves, grates, or steam boilers, than dry wood. If exposed in the open air it is apt to crumble, but if protected it receives no special injury. Dr. Torrey thinks there is no reason why it should not be eminently useful for generating steam and for smelting ores.

Throughout the intercalated beds of clay at Boulder Creek and vicinity are found masses of a kind of concretionary iron ore, varying in size from one ounce to several tons in weight. This iron ore is probably a limonite commonly known under the name of brown hematite or brown iron ore. It may perhaps be found in the state of carbonate of iron when sought for, beyond the reach of the atmosphere. These nodules or concretionary masses, when broken, show regular concentric rings varying in color from yellow to brown, looking sometimes like rusty yellow agates. It is said to yield seventy per cent. of metallic iron. The first smelting furnace ever erected in Colorado was established here by Mr. Marshall, and he informed me that for the production of one ton of pig iron, three tons of the ore, two hundred pounds of limestone, and one hundred and thirty to one hundred and fifty bushels of charcoal are required. Over five hundred tons of this ore have been taken from this locality, and the area over which it seems to abound cannot be less than fifty square miles. Indications of large deposits of iron ore have been found in many other localities along the line of the Pacific railroads, and if the mineral fuel which is found here in such great abundance can be made useful for smelting purposes, these lignite and iron ore beds will exert the same kind of influence over the progress of the great West that Pennsylvania exerts over all the contiguous States. When we reflect that we have from ten thousand to twenty thousand square miles of mineral fuel in the center of a region where for a radius of six hundred to one thousand miles in every direction there is little or no fuel either on or beneath the surface, the future value of these deposits cannot be overestimated.

The geological age of these western lignite deposits is undoubtedly tertiary. Those on the Upper Missouri have been shown to be of that age both from vegetable and animal remains, and in the Laramie Plains I collected two species of plants, a *Populus* and a *Plantanus*, specifically identical with those found on the Upper Missouri. The simple fact that cretaceous formations Nos. 1, 2, 3, 4, and 5, are well shown all along the foot of the mountains, and that No. 5 presents its usual lithological character with its peculiar fossils, within fifteen miles of Marshall's mines, also that at the mine, 2, 3, and 4 are seen inclining at nearly the same angle and holding a lower position than the lignite beds, is sufficient evidence that the strata inclosing the lignite beds are newer than cretaceous. A few obscure dicotyledonous leaves were found, which belong rather to tertiary forms than cretaceous.

The connection of the lignite deposits on the Upper Missouri has been traced uninterruptedly to the North Platte, about eighty miles above Fort Laramie.

They then pass beneath the White River tertiary beds, but reappear again about twenty miles south of Pole Creek, and continue far southward into New Mexico. Near Red Buttes, on the North Platte, it seems also probable that the same basin continues northward along the slope of the Rocky Mountains nearly or quite to the Arctic Sea. Whether or not there are any indications of this formation over the eastern range in the British possessions, I have no means of ascertaining, but the Wind River chain, which forms the main divide of the Rocky Mountain Range, exhibits a great thickness of the lignite tertiary beds on both eastern and western slopes, showing conclusively by the fracture and inclination of the strata, that prior to the elevation of this range, they extended uninterruptedly in a horizontal position across the area now occupied by the Wind River chain. Passing the first range of mountains in the Laramie Plains, we find that the Big Laramie River cuts through cretaceous beds, Nos. 2 and 3, continuing our course westward to Little Laramie, a branch of the Big Laramie, and No. 3 becomes fifty to one hundred and fifty feet in thickness filled with fossils, *Ostrea congesta*, and a species of *Inoceramus*. At Rock Creek, about forty miles west of Big Laramie River, the lignite beds overlap the cretaceous, but in such a way as to show that the more inclined portions have been swept away by erosion, and that the red beds and carboniferous limestones once existed without break and in a horizontal position across the Laramie Range prior to its elevation.

I cannot discuss this matter in detail in this article, but the evidence is clear to me now, that all the lignite tertiary beds of the West are but fragments of one great basin, interrupted here and there by the upheaval of mountain chains or concealed by the deposition of newer formations.

When I wrote the article on the lignites of the West, all my own investigations pointed strongly to the conclusion that no coal beds of any great value, in an economical point of view, would ever be found in the West in formations older than the tertiary. When my large collections of vegetable and animal remains from the coal beds in Wyoming, Colorado, and New Mexico, now deposited in the Smithsonian Institution, are carefully studied, I can speak with more confidence on that point. I can say just here that I have as yet seen no reason to change that opinion so far as my own observations are concerned.

In the spring of 1868, Professor Lesquereux, who is so justly celebrated for his skill in the study of fossil plants, sent me the following valuable notes as the result of a preliminary examination of some leaf impressions from the coal deposits in various parts of the West. His conclusions seem to confirm my opinions that all these coal formations are of tertiary age.

SPECIES FROM ROCK CREEK, LARAMIE PLAINS.

1. *Populus attenuata*, Al. Braun. The identity of these leaves with the European species is undoubted.
2. *Populus lævigata*, sp. nov., related to *P. balsamoides*, Göpp., a species which, like the former, is abundant in the miocene of Europe.

3. *Populus subrotunda*, sp. nov. Type of neuration of *P. melanaria*, Heer, and form of leaves of *P. mutabilis*, Heer, both species also common in the miocene of Europe.

4. *Quercus acrodon*, sp. nov., a fine oval leaf resembling a chestnut leaf, related to *Quercus prinoides*, Wild, of our time.

5. *Quercus haydeni*, sp. nov., lyrate leaf with lobes strongly dentate, without near relation to any species either of the tertiary or of our time.

6. *Platanus, aceroides*, Göpp., one of the most common species of the miocene of Europe. It is closely related to, if not identical with, *P. occidentalis*, *L.*, of our time.

MARSHALL'S MINE (NEAR DENVER.)

1. *Quercus chlorophylla*, Ung. Three specimens of this species have been figured and described in my paper, "On species of fossil plants from the tertiary of Mississippi," (Trans. Phil. Soc., vol. 13, pl. xvii, figs. 5, 6, 7.) It is still uncertain if these leaves represent a *quercus*, but all belong to the species described and figured by Heer under this name, and common in the whole thickness of the European miocene.

2. *Quercus lyelli*, Heer, also figured in the above paper, pl. xvii, figs. 1, 2, 3. Though the specimen is somewhat obscure, the essential characters which distinguish the species are well discernible. It is abundant in the Bovey Tracy lignite formations of England, lower miocene.

3. *Cinamomum affine*, sp. nov. This species is also found at Raton Pass. The leaf from Raton Pass is smaller and might belong to a different species, but except the size I do not find ground for separation; very near *C. mississippiensis*, Lesq., and also closely related to *C. buchi*, Heer, of the lower miocene of Europe.

4. *Cornus incompletus*, sp. nov. A part of a leaf apparently round at the top, general outline uncertain. It is figured merely for future reference. By its peculiar nervation this leaf appears in close relation to, if not identical with, *Cornus rhamnifolius*, Web. Pretty common in the lower miocene of Europe.

5. There are in the Marshall's shales a few fragments of maple leaves (*acer*) specifically undeterminable, and also one winged seed of this genus. This seed has a narrow straight wing like that of *Acer trilobatum*, Heer, but with smaller nutlet.

6. *Rhamnus salicifolia*, sp. nov., in soft sandstone; related to *R. marginatus*, Lesq., and and also to *R. carolinianus*, Walt., now living and abundant in southern swamps.

7. *Juglans rugosus*, sp. nov., very nearly related to *J. acuminata*, Al. Braun, a species extensively distributed in the European miocene.

8. *Echitonium sophiæ*, Web. The leaf has no visible nervation, but it is exactly like both the forms represented from European specimens. It is found in the whole miocene of Europe, especially in the lower stage.

9. *Phyllites sulcatus*, sp. nov. The borders of the leaf are destroyed, but the nervation is quite peculiar. It is referable either to a *Rhodora* like *R. canadensis* of our time, or represents merely the lower part of the winged petiole of the fruit of a linden, (*Tilia.*)

10. *Lygodium compactum*, sp. nov. Though many species of lygodiums are described from the tertiary of Europe, none are related to ours. One lobe of a leaf only is presented, and the general outline of the leaf is therefore unknown, but the nervation, which is very close and more like that of a *Neuropteris*, is of a peculiar character.

LIGNITE BEDS NEAR GOLDEN CITY, COLORADO.

1. *Magnolia tenuinervis*, sp. nov. Not possible to indicate the general form of the leaf of which a part only is presented. Its thin and sharp secondary nerves distinguish it from any other fossil species.

2. *Lathræa arguta*, sp. nov. May be a *Pecopteris*. No relation observed of any known species to this one.

RATON PASS. SPECIMENS COLLECTED BY DR. LECONTE.

1. *Berchemia parvifolia*, sp. nov. Related to *B. multinervis* of the European miocene, but still more like our *Berchemia volubilis* which fills the southern swamps. The basilar part of the leaf is not seen and therefore a satisfactory determination is not possible.

2. *Abietites dubius*, sp. nov.

Most of the specimens from Raton Pass have some remains of leaves or branches of a coniferous species which can be referred, perhaps, as well to the genus *Araucaria* as to *Sequoia* or *Abies*. As the leaves on the branchlets appear evidently placed around the stems and not on both sides of it, and as the scars left on the bark are of the same form as those of an *Abies*, I place these remains in this genus till they may be studied on better specimens. The leaves are pointed as in *Taxites dubius*, Göpp., from the tertiary of Europe; except this, these remains have no analogy with any other, published or figured.

3. *Echitonium sophiæ*, Web. A small fragment exactly like those of Marshall's coal bed and a specimen of *Cinnamomum affine*, already mentioned, from the Marshall's shales.

UPPER END OF PURGATORY CAÑON, DR. LECONTE.

1. *Rhamnus obovatus*, sp. nov. All the specimens are from the same place, and all contain fragments of the same species, and none of any other. This species is peculiar by the form of the leaves; it has the character of a *Rhamnus* but the secondary nerves are closer and more numerous than in any other species of the genus, even more so than in a *Berchemia*. I do not know of any fossil plant comparable to this.

From this short report on your fossil plants examined till now, it is easy to draw some general conclusions.

From Rock Creek we have only six species. Two are identical with species from the miocene of Europe, and one of them, *Platanus aceroides*, is not distinguishable from our *P. occidentalis*. Two other species are closely allied to European tertiary species. And of the two others, one is an American type related to *Quercus prinoides*, still in our flora, the other a peculiar and lost type. The appearance of this florula is quite modern. This may be the result of geographical circumstances. Poplars and buttonwoods live together in the bottoms of rivers, and therefore I may mistake in believing this Rock Creek formation more recent than that of Marshall's. In any case it is certainly tertiary and has no plants of an older formation.

In Marshall's (coal beds) we find only ten species of fossil plants, two *Quercus* and one *Echitonium* apparently identical with miocene species of Europe, one *Rhamnus*, closely related to a living species of ours, and at the same time to a fossil species of the lignite of Mississippi, one *Cornus*, one *Juglans*, and one *Cinnamomum*, all related to miocene species, and the last one also closely allied to a species of the Mississippi tertiary; undeterminable leaves of maple, seeds of the same genus, a *Lygodium* and an undeterminable *Phyllites* complete the list. These plants have, therefore, all of them, the character of tertiary plants. The general aspect of the Marshall coal flora is that of the Mississippi lignite, which I consider as either lowest miocene or eocene. In this I am much pleased to find my views so well agreeing with yours.

The materials obtained from the strata of Golden City, Raton Pass, and Purgatory Cañon, are too scanty to permit considerations in regard to the geological positions of the strata which have furnished them. No *Abies* has yet been described from tertiary strata, but with these broken remains of a conifer of uncertain genus, the shale of Raton Pass has a *Berchemia*, which is a tertiary plant, and a leaf of *Echitonium*, and one of *Cinnamomum* identical with specimens found at Marshall's.

In conclusion, I beg leave to say, that while I have the most profound respect for the labors of my fellow geologists in the same field, I differ with them somewhat, simply because the evidence, to my mind, points in a different direction. In various portions of the Laramie Plains, Colorado, Raton Hills, &c., I have observed between the well-defined cretaceous and tertiary beds a group of strata composed of thin layers of clay, with yellow and gray sands and sandstones, which I have called transition or beds of passage. If in these beds I were to find some purely marine remains, even inoceramus or baculites, I should then call them transition beds, in accordance with the evidence of the continuous uninterrupted growth of the continent from the cretaceous through the tertiary period. There is no proof, so far as I have observed, in all the western country of true non-conformity between the cretaceous and lower tertiary beds, and no evidence of any change in sediments or any catastrophe sufficient to account for the sudden and apparently complete destruction of organic life at the close of the cretaceous period. In all my examinations of the coal formations over so vast an area, I have never yet seen a trace of a cretaceous fossil in any strata above the coal. One of the most important practical questions for solution in the west is, whether these coals can be rendered useful for smelting ores. To aid in the solution of this question, I have appended the following analyses of the coals from various portions of the West.

Mr. J. P. Carson, my assistant on the United States geological survey, 1868, made the following analysis of a fair specimen of the coal from the Carbon mines, Northern Pacific railroad.

Moisture at 100° F	11.60
Volatile combustible matter	27.68
Fixed carbon	51.67

Ash	6.17
Sulphur	2.88

Color of ash, light gray. Specific gravity, 1.37. Weight per cubic yard, 2,212 pounds.

My assistant, Persifor Frazer, jr., in the United States geological survey of Colorado and New Mexico during the past season, has made the following analyses of coals along the line of the Union Pacific railroad. They were made with great care and I have the most perfect confidence in their accuracy:

Coal from mine at Point of Rocks:

	Per cent.
Carbon	64.70
Ash	4.40
Sulphur	0.42
Water and volatile substances	30.48
Total	100.00

Coal from Rock Creek:

	Per cent.
Carbon	61.34
Sulphur	2.00
Ash	1.50
Volatile substances and water	35.16
Total	100.00

Coal from Black Buttes:

	Per cent.
Carbon	71.64
Sulphur	2.00
Ash	2.50
Volatile substances and water	23.86
Total	100.00

Coal from the Evanston mine was tested for its carbon alone and found to contain carbon, 72.16 per cent. All these coals resemble in their physical properties those met with along the route of the Colorado and New Mexico survey.

I take the liberty of quoting in this connection the following analyses of coals from the admirable report* of my friend Doctor J. H. LeConte. I found this report, as well as that of Doctor Newberry, of great service to me in my explorations during the past season:

Locality.	Fixed carbon.	Volatile material.	Water.	Ash.
1. NEW MEXICO.				
Vermejo Cañon	59.72	23.73	3.27	13.28
Placer anthracite	88.91	3.18	2.90	5.21
2. COLORADO.				
Murphy's, near Denver	55.31	29.07	11.70	3.92
Marshall's, near Denver	59.20	26.00	12.00	2.80
Coal Creek†	57.70	19.30	20.00	2.00

*Notes on the geology of the survey for the extension of the Union Pacific Railway, eastern division, from the Smoky Hill River, Kansas, to the Rio Grande, by John L. LeConte, page 58.

†The analyses of the Marshall coal and that of Coal Creek were made by Doctor Torrey, and are copied from Doctor Hayden's paper in Silliman's Journal for March, 1868

Locality.	Fixed carbon.	Volatile material.	Water.	Ash.
3. PACIFIC COAST.				
A.—Cretaceous.				
Bellingham Bay, Washington Territory	45. 69	33. 26	8. 39	12. 66
Nanaimo, Vancouver's Island	46. 31	32. 16	2. 98	18. 55
B.—Tertiary.				
Coos Bay, Oregon	41. 98	32. 59	20. 09	5. 34
Mount Diablo, California	40. 65	40. 36	13. 47	5. 52
Do	46. 84	33. 89	14. 69	4. 68
Do	44. 92	40. 27	13. 84	0. 97
Do	44. 55	37. 38	14. 13	3. 94
Do	36. 35	35. 62	20. 53	7. 50

German tertiary coals.

Variety.	Carbon.	Hydrogen.	Combined water.	Hygroscopic water.
Fibrous, (faserige)	48	1	31	20
Earthy, (erdige)	56	2	22	20
Laminated, (muschlige)	60	3	17	20

The ash is neglected in the foreign analyses, but is stated to average from 5 to 10 per cent. When first mined, the German brown coals contain frequently nearly 50 per cent. of hygroscopic water, which by drying is reduced to 20 or 25 per cent.

The absolute heat effects of the German coals are given as follows:

Variety.	Air-dried.	Kiln-dried.
Fibrous	.50	.63
Earthy	.62	.76
Laminated	.70	.84

The data obtained by Professor Brush by the reduction of oxide of lead, when placed in a decimal form, pure carbon being unity, are:

Vermejo Cañon .67
Placer anthracite .91
Denver, (Murphy's) .60

The following are analyses of water from springs, &c., by Mr. P. Frazer, chemist and mineralogist to the United States geological survey of Colorado and New Mexico:

While in Rawling's Springs I was employed by the Union Pacific Railroad Company to examine the waters from various springs, which incrusted the boilers of locomotives and stationary engines of the company, as well as of coals from the principal coal-beds on the line of the road. The result of these analyses I append:

Scale from the boiler of an engine in the machine shop at Rawling's Springs.—This scale was of a dark color due to impurities in suspension in the water. It consisted of the chlorides of potassium and sodium, the sulphates of lime and magnesia and the silicate of alumina. The major part of the soluble matter was composed of salt and gypsum. Some water from a salt pond in the Black Hills, some distance from Sherman, was analyzed and found to contain chloride of sodium, chloride of potassium, the carbonate of soda, and some alumina.

Boiler scale from locomotive running between Rawling's Springs and Bryan.—This scale was of a gray color, but proved to be of the same chemical constitution as that previously given, viz, chlorides of potassium and sodium, sulphates of lime and magnesia, and the silicate of alumina.

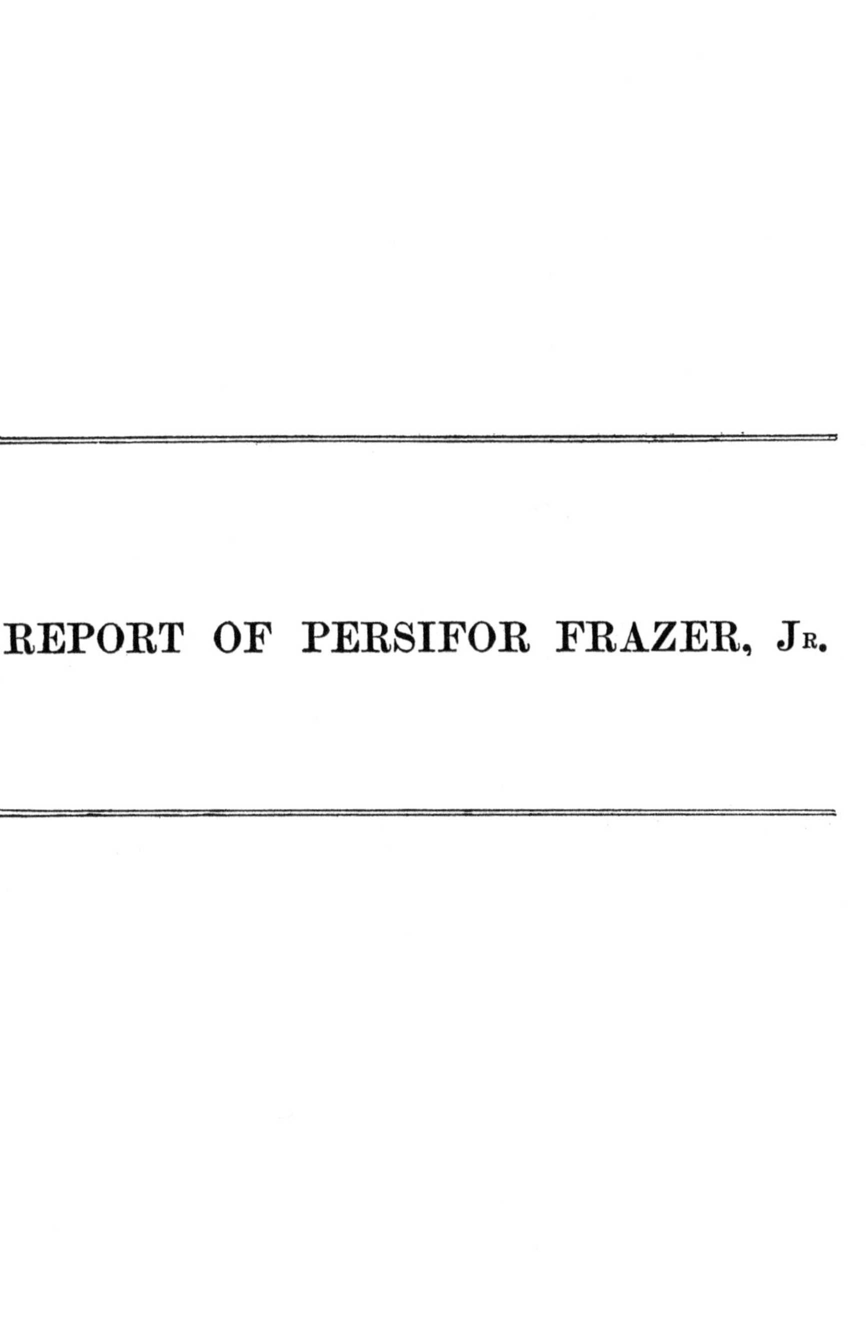

REPORT OF PERSIFOR FRAZER, Jr.

MINES AND MINERALS OF COLORADO.

DENVER, COLORADO, *October* 15, 1869.

SIR: I have the honor to report that the examination of the minerals, and the means employed to utilize them, in the Territories of Colorado and New Mexico, which you directed me to make, has been conducted as well as the very limited time at my disposal would permit, and a preliminary report of the results is herewith respectfully submitted.

In the letter accompanying the first report to the Secretary of the Treasury by the commissioner appointed to collect the same kind of information from the country lying west of the Rocky Mountains, Mr. Browne urges that the six months which were prior to the meeting of Congress would not permit of any but a most imperfect treatment of the subject, and limits himself to sketching an outline of the work to *be* done.

The same is true in a much greater degree of the few weeks in which I was obliged to gather the materials for this report, especially as the greater portion of the time was spent on the march, remote from all points where statistics were accessible.

Any report of the condition of mining affairs in the Territories of Colorado and New Mexico, (each of which is larger than all the New England States put together,) and in particular of the former, which counts its discovered lodes, the varieties of its minerals, and its mining enterprises, by thousands, and in which energetic capital and intelligence, "ever striving through darkness to the light," are working such incessant changes, must represent things as a telescope represents the stars, not as they are or ever were, but this as it was last week and that as it was last year.

In consideration of these difficulties, I venture to hope that you may regard all shortcomings more leniently, and that the following, though far from complete, may not altogether fail to answer the requirements of Congress.

In conclusion, I would call attention to the great courtesy and kindness I have experienced in the course of my investigations from the citizens of the two Territories generally, the owners and superintendents of the various mines and mills, the possessors of cabinets of minerals, &c., and the officers and their families stationed at Forts Union and Garland.

Especially do I thank Mr. J. Alden Smith, the mining editor of the Central City Register; Mr. D. J. Ball, of Empire City; Colonel Anderson, of the Real Dolores; and Mr. Cheever, of the Brown Mining Company in Georgetown, for the assistance, in a professional way, which they have rendered me; nor can I forget the kindness of Mr. Marshall, of Black Hawk, and Mr. Schultz, of Central City.

Where it was not possible for me personally to inspect the mines of which I have spoken, I have in every case stated that the information is given on the authority of others.

I remain, sir, with great respect,

PERSIFOR FRAZER, JR.,
Mining Engineer.

DR. F. V. HAYDEN, *United States Geologist.*

A natural division of the subject about which information has been sought would seem to be; I, the minerals, and II, the mines of Colorado and New Mexico; and these again into I. 1, the minerals of commercial value, and I. 2, those of no commercial value, but more or less characteristic of the rocks or formations in which they occur.

The mining portion of this report would have been better divided into II. 1, gulch or placer mining, and II. 2, lode or legitimate mining, while under the latter head the subject would naturally divide itself into *a*, the methods in use for getting out the ore and taking care of the mines; *b*, the dressing of the ores by mechanical processes; and *c*, the chemical treatment of the ores, their reduction and preparation for the market, or shipment out of the Territory. This would be a natural division of the subject, but the time, and, consequently, the opportunities of observation have been so insufficient for the above thorough treatment of the subject that I have deemed it better to forward to you, as my part of the preliminary report, only the notes I have made in the field, with a few observations on various points connected with the subject.

In a belt, of which it would be difficult to define the limits, but which may be generally stated as lying east and west of the great continental divide as far as the gneiss or granite extends, and reaching north and south as far as investigation has made the Rocky Mountain chain known to us, lie the ores of the precious, and some of the baser, metals. Of the distribution of this great mineral wealth throughout the hundreds of leagues of this belt very little is known, the small area which has become the prize of the gold-seeker furnishing wholly insufficient data upon which to base general conclusions.

To begin with, the rock in which occur all these lodes is that which carries the precious metals, with rare exceptions, the world over, and which is either a granite or a gneiss, or, as in the Central City district, such an inextricably confused mixture of both that it were impossible to call it either. This is the country rock. Whether from the great changes to which this rock has been exposed through countless ages, or whether from other causes, it shows itself in most various forms at different places, and passes by imperceptible phases through gneiss, granite, syenite, and porphyry. This porphyry is perhaps more frequently observed in the neighborhood of veins.

A fine illustration of the irregularity with which these rocks succeed each other is to be observed along the road from Mount Vernon through Idaho City to Georgetown. Along Clear Creek, from Fall River to Georgetown, the inclination and direction of the rocks appear to be as variable as their structural character, a general northwesterly dip being perhaps most common, while red and gray, heavy-bedded, and thinly-laminated gneiss and red and gray granite succeed each other in utter confusion. Here and there a vein of quartz or quartz-porphyry or syenite (very frequently auriferous) is visible, forming a light-colored streak usually down the sides of the opposite hills. This composite character of the country rock has been noticed, as I am informed, in most, if not all, of the mining districts, and on both sides of the Sierra Madre or main range. The gangue rock is most frequently quartz, which, of course, assumes very different appearances at different places, both in texture and in color. In some cases the gangue rock is porphyry more or less weathered. (Brown Lode, West Argentine, *et al.*)

The minerals of Colorado of commercial value which are most widely distributed are auriferous iron and copper pyrites, (malachite and the sulphates of iron and copper from their decomposition, though nowhere in large quantities, being spread over wide areas,) zincblende, argentiferous galena, brittle silver ore, fahlerz, specular iron, hematite and magnetic pyrites, cerussite and anglesite, native gold and silver, horn silver, embolite, (confined chiefly to the neighborhoods of Georgetown and Snake River, I believe, as far as yet ascertained,) titanic iron ore, micaceous iron ore, spathic iron ore, Smithsonite, copper glance, coal and Albertine coal. These comprise the principal ores which I have observed, but time and more thorough search will undoubtedly disclose to the mineralogist, if not to the metallurgist and miner, many as yet hidden treasures.

Gilpin County and the region about Empire are rated as gold fields, and the values of ores from these and some other districts are given in ounces of gold per ton; whereas the adjacent country around Georgetown, abounding as it does in argentiferous galena and silver glance, (called simply "sulphuret,") has the number of ounces silver per ton as its standard. In some few veins, as the Whale Lode near Idaho City, the values of the gold and silver present in the ore are nearly equal.

A more detailed specification of the ores follows:

Iron pyrites, (FeS_2.)—Almost universal in the mines. Occurs in cubes from the size of a pin's head to those of an inch on the sides. Also in pentagonal dodecahedra.

Copper pyrites, (Cu_2S+FeS_2.)—Is only second to iron pyrites in the frequency of its occurrence.*

Zincblende, (ZnS.)—Is also very common, especially in the Georgetown region. Fine specimens were obtained from the Baker Lode, West Argentine and the Griffith Lode, close by Georgetown. Also from Gilbert's (formerly Commonwealth Mining Company) Lode, near Nevada City.

Galena, (PbS.)—Usually argentiferous. In all the lodes in the vicinity of Georgetown. Contains from one hundred to six hundred ounces silver per ton.†

Brittle silver ore, (Stephanite $5A_gS+Sb_2S_3$.)—Occurs in the silver mines of Georgetown. (Terrible and Brown lodes.)

Fahlerz, [$(4RS+4Cu_2S)\,QS_3$. R=Fe, Cu, Zn and often some Ag and Hg = Q=Sb and As.]—Also in the region around Georgetown. The formulæ here given are from Naumann's Mineralogy. I am not aware that Hg has been discovered in this ore, but as it coincides in its physical properties with the ordinary fahlerz, I append the above formula.

Light ruby silver, (*Proustite*,) ($3AgS.AsS_3$); *Dark ruby silver*, (*Pyrargyrite*,) ($3AgS.SbS_3$.)—Handsome specimens of these two ores were observed intermixed with the galena from the Brown Lode. Also from Snake River.

Silver glance, (AgS.)—From the Georgetown neighborhood. Equator and Terrible lodes. A ton of galena, containing much of this ore, was recently sold by a gentleman of Central City to Professor Hill for $1,900 cash, and the latter realized a profit of $700 from it.

---

* Both iron and copper pyrites of this region contain gold in indefinitely fine particles; The former is, in fact, *the* gold ore. Where these minerals have been exposed to the action of the weather, they have been decomposed and the gold set free. The value of the gold in a ton varies from nothing to five hundred dollars, and even more. I have observed small octahedra of gold on the crystal faces of iron pyrites from the Pleasant View mine near Central City.

† It is somewhat remarkable that these veins of galena generally "pinch up" or grow smaller as the depth increases. I take this general statement from the best authority I could obtain on the subject. A gentleman well acquainted with the Georgetown ores informed me that all attempts hitherto to produce lead for the market had failed from deficiency in the supply of galena. This statement, which I give for what it is worth, appears all the more remarkable when one compares it with the experience of miners in Freiberg, Przibram, and Clausthal.

Copper glance. (Cu_2S.)—Bergen district, near Idaho City, Pleasant View, &c.

Malachite, ($CuO . CO_2$;) *Blue vitriol*, ($CuO . SO_3+5HO$;) *Green vitriol*, ($CuO . SO_3+7HO$.)—Occur in various mines from the decomposition of the pyrites.

Pyromorphite, ($PbO . PO_5+PbCl$.)—Associated with the galena of various mines near the surface.

Specular iron ore, ($FeO . Fe_2O_3$.)—Cache à la Poudre, St. Vrain's, &c.

Red and brown hematite, (Fe_2O_3 and Fe_2O_3+HO.)—Of frequent occurrence in the vicinity of the coal.

Coal.—Beds of coal occur all along the flanks of the mountains, but in the property of Mr. Marshall are perhaps the best exposures. Here are no less than nine outcrops. They make their appearance at various points along the range as far down as Santa Fé, and are of unknown extent. Albertine coal, or solidified petroleum, is stated by Prof. Denton to occur on White River, in the western part of the Territory.

Gold.—Occurs in the neighborhood of Central City, in the German lode, and many others. In the Placer diggings. Some beautiful crystals attached to cubes of iron pyrites in the ore from the Pleasant View mine.

Silver.—In many mines as wire or hair silver, Brown and United States Coin lodes.

Cerussite, ($PbO. CO_2$.)— Pleasant-View mine.—In small translucent crystals occurring in geodes.

Anglesite, ($PbO . SO_3$.)—Freedland lode, Trail Run.

Horn silver, ($Ag Cl$.)—Georgetown, Snake River.

Embolite, ($AgBr+AgCl$.)—Peru district, Snake River.

Titanic iron ore, ($x\ Ti_2O_3+y\ Fe_2O_3$.)—Quartz Hill, and Russel Gulch, near Central City.

Micaceous iron ore, (Fe_2O_3.)—Elk Creek. In fine crystals like mica.

Spathic iron ore, ($FeO . CO_2$.)—Eureka and Griffith lodes, &c.

Smithsonite, ($ZnO . CO_2$.)—Running lode, Blackhawk, &c.

Salt, ($NaCl$.)—From Salt Springs in South Park, twenty miles south-east of Fairplay. Can produce forty thousand pounds *per diem*.

By characteristic minerals, I mean to include all those that have no commercial value. They furnish proof, in most cases, of the presence of other minerals, of rocks or of formations. Of the characteristic minerals, among the most common are—

Hydrated oxide of iron, (brown ochre, yellow ochre, bog iron ore, &c.)—Occurs with the coal beds at South Boulder, Golden City, &c., &c., and is frequently regarded as a surface indication of the presence of gold, silver and the precious ores generally.

Quartz, (SiO_3.)—The most important of the characteristic minerals. Very widely diffused. Forms the gangue of nearly all the veins of the precious metals in Colorado. As gangue rock it crops out on the hill sides in white or colored streaks, usually intersecting the planes of stratification of the rocks. Uncrystallized, presenting sharp and jagged edges, and a broken conchoidal uneven fracture, sometimes weathered by the disintegration of the minerals it contained. Pebbles and partially rounded crystals of quartz are abundant in the prairies east of the Rocky Mountains, whence they have been carried down, and may be observed hundreds of miles east of the easternmost "hog-back." Indeed, the abundance of these small pebbles of quartz and of the red feldspar is very remarkable, occurring as they do in great quantities on the summits of the little prairie hillocks at such an immense distance from their place of origin.

1. Smoky quartz and black quartz.—Elk Creek.
2. Rock or Berg crystal.—Near "Dirtywoman's Ranch," and in geodes in various mines.
3. Rose quartz.—Quartz Hill.
4. Agate, (moss agate, &c.)—Middle Park, Arkansas River Park, &c.
5. Amethyst.—Nevada City, Mill City, &c.
6. Aventurine.—Elk Creek.
7. Heliotrope, (bloodstone.)—Middle Park.
9. Carnelian.—South Park and Middle Park.
10. Chalcedony.—South Park, Trout Creek Pass, &c.
11. Chrysoprase.—Middle Park.
12. Jasper.—South and Middle Parks.
13. Onyx.—Middle Park, Grand River, &c.
14. Sardonyx.—Golden City, Mount Vernon.

Hornstone, flint, milk quartz, prase, catseye, firestone, and other different varieties of silicic acid, are met with in the above localities, but have no especial interest.

Opal, (hydrated silicic acid.)—Idaho City, Golden City, South Boulder, &c.

Feldspar.—Very abundant in the mountains and as boulders and pebbles throughout the Territory. Associated with quartz in the granites, gneisses, and porphyries of the gold-bearing mountains.

a. Orthoclase ($Al_2O_3 . 3SiO_3 + KO.SiO_3$) is largely the predominant feldspar in the rocks of Colorado.

a 1. Pegmatolite.—Flesh-red, orthoclastic, abundant as pebbles, scattered with quartz over the prairies for hundreds of miles. Forms red granites and gneisses with quartz and mica, and red syenites with hornblende. Very common.

a 2. Adularia.—Forms a white porphyry when associated with quartz in many places along Fall River, and in many veins. Not common.

a 3. *Sanidin*.—Fine crystals of hopper-shaped sanidin from Quartz Hill.

b. Plagioclastic feldspars.

b 1. Albite, ($Al_2O_3 . {}_3SiO_3 + NaO . SiO_3$.)—Trout Creek Pass.

b 2. Oligoclase.—Arkansas River Park, &c.

b 3. Labrador, ($Al_2O_3 . SiO_3 + CaO . SiO_3$.)—In the basalts and diabases of the region about the Spanish Peaks, Trinidad, the upper part of San Luis Park, and the Puntia Pass.

Hornblende, (silicate of lime, magnesia, and suboxide of iron).—In the syenite in and around Idaho.

Diorite.—Near Empire City and elsewhere.

Garnet.—South Park, twenty miles from Fairplay. Breckenridge.

Mica, ($KO . SiO_3 + Al_2O_3 . SiO_3 + RO . SiO_3$.)

1. Potash mica.—Light colored. Frequent in the gneisses of Gilpin and other counties.
2. Magnesian mica.—Dark colored. Frequent in the gneisses of South Park, Trout Creek Pass, &c.

Leucite, ($Al_2O_3 .. SiO_3 + KO . SiO_3$.)—In trachytic lava between the Cuchara and the Apishpa.

Chlorite.—In diabase, near Trinidad.

Amphibole, (augite).—In basalts, near Trinidad, and diabase near Apishpa.

Epidote, ($CaO . SiO_3 + [Al_2O_3 + Fe_2O_3] SiO_3$.)—Trail Creek.

Tourmaline.—Guy Hill.

Calc spar, ($CaO . Co_2$.)—Very widely distributed. Idaho, &c., &c.

Gypsum, ($CaO . SO_3 + HO$.)—Interstratified in the new red sandstone or triassic beds. South Park, &c. Also, accompanying the coal in thin scales.

Anhydrite.—Elk Creek.

Salt, (NaCl.)—In solution in many springs. As deposit on rocks in their vicinity.

Heavy spar, ($BaO . SO_3$.)—As gangue rock in many mines. Baker lode, &c.

Meteoric iron.—Found near Bear Creek.

Beryl, ($Al_2O_3 . 2SiO_3 + Gl_2O_3\ 2SiO_3$.)—Bear Creek.

Brucite, (MgO . HO.)—James Creek.

Idocrase, [$(CaO + MgO)\ SiO_3$]—Bear Creek.

LAPORTE ON THE CACHE À LA POUDRE.

The town lies on both banks of the above creek. The appearance of the country is that of a number of superposed layers or strata dipping from the mountains, and presenting a steep and more or less rugged basset face toward them. The cañon along which the river makes its way through these "hog-backs" intersects the latter nearly at right angles. We followed a cañon to the north of that of the river, and rode twelve miles to the extremity of the bluff on the left. The bluff to the right hand was broken, and exhibited a clearly defined stratified side with red sandstone, limestone, and conglomerate succeeding each other in the order named.

On turning the extremity of the bluff to the right we came upon a very weathered syenite region remarkable for the redness of its talus.

The mineral veins which our guide brought us to see were all situated within an area of a square mile or so, in these syenite hills.

The first proved to be a dike or vein of syenite intersecting an older rock of the same, which showed on the surface a very thoroughly decomposed rock, containing an excess of iron, which gave it a specific gravity rather higher than usual and a dark brown color. Hornblende predominated in the rock. There were here and there traces of various ores of copper, and lining the walls of the small cavities in the rock was observed a thin film of gypsum and chloride of sodium. This spot was situated upon the east bank of the North Fork of the Cache à la Poudre, and about three hundred feet above that stream.

The next opening we visited was about half a mile northwest, and was called Maxwell's opening. This was again a dark-colored, not very distinguishable syenite, coated with malachite, and more or less permeated by copper pyrites. The opening was seven feet deep and the crevice four or five feet wide, and the two pay streaks situated, the one against the south wall rock, and the other about thirty inches therefrom. The ore becomes harder and more solid the deeper it is found.

Hole No. 3 was three hundred yards from No. 2. It was about four feet deep, two and a half inches wide, and four feet long. The rock was silicious and intimately mixed with a yellowish clay, which, with the reddish tinge due to the oxide of iron, gave the whole mass a copper color, which probably misled the prospectors and caused the digging of the hole. A little copper pyrites was observable and a very little malachite.

Hole No. 4 had been sunk by some Frenchmen fifteen feet deep, three and a half feet wide, and five feet long. The rock described as composing No. 3 occurs with a curious slag-like silex containing very plain pseudomorphs of cubes of iron pyrites. In this ore was a little copper pyrites and malachite.

Lastly we emerged upon a precipitous narrow dike of quartz porphyry overhanging the before-mentioned creek on its right bank, and forming an abrupt wall of one hundred and fifty feet above its bed. The crystals, both of quartz and feldspar, were very large, averaging the size of a man's hand. The quartz was standing transparent and milky, while the feldspar was of the true flesh-red color common to typical pegmatolite.

BOULDER CITY, JULY 5.

Obtained specimens of ores from leads ten miles from this place up the James Creek. A fine solid specimen of argentiferous galena was given to me from one foot beneath the surface at the intersection of the Buckhorn and Big Thing lodes. Mr. Arnett, the owner of the claim, states that this ore runs from $125 to $200 per ton in silver, and $300 in gold. I also obtained a specimen of very fine-looking ore from the Horsefall mine, ten miles from Boulder City, in Gold Hill.

Near Boulder City, on the property owned by Mr. Marshall, occur some fine exposures of coal, which have been visited by Dr. J. LeConte, and examined subsequently with much care by yourself, so that a special report from me upon them would be superfluous. I will confine myself, therefore, to the mere statement that, in a distance east and west of a couple of miles, there are eleven exposures of very excellent coal, at least nine of which would seem to promise rich rewards for the working. The mining which has as yet been done, was merely to fix the location and investigate the extent of the veins, as well as that could be done at the surface. The beds appear to be large enough to yield with proper appliances a thousand tons a day for an indefinite time. The commercial value of this coal when the country is a little more settled can hardly be overestimated. The color is a dark brownish or bluish black, with a high luster and low specific gravity. It breaks, as does all of this recent coal which I have observed along the flanks of the Rocky Mountains, with the exception of the rare anthracite, into parallelopipeda. This friability is annoying to the smelter, who finds that it chokes up his grate bars and stops the draught, but it has been successfully combated in the works of Professor Hill, of Blackhawk, by the use of the staircase furnace. This coal contains very few impurities, and can be and is used in the blacksmith's forge without previous coking. Specimens have been procured from these various veins and will be analyzed at the earliest opportunity and the results submitted to you.*

GOLDEN CITY.

Golden City is situated nearly west from Denver, on a gently sloping plain at the inner extremity of the cañon between two singular mesas or table mountains of igneous rock, capped, like the innumerable mesas further south, with thick slabs of basalt. The western border of the beautiful valley in which Golden City is built, is formed of the gneissic rocks, upon which rest the triassic (partly variegated and partly white) beds, and then follow the jurassic and cretaceous, but ill-defined on account of the unbroken grassy sward which usually conceals them. The dip of the tertiary beds is here beyond the vertical, so that they seem to incline toward the mountains. There is a lead of silica in a state of fine division which has been opened on a hill of triassic. On the west side of

*I forward to you as a supplement to this report analyses of some coals from Wyoming Territory, and hope to add the Boulder coals thereto shortly.

a little valley separating the cretaceous and tertiary, occur, in order, clay, sandstone, clay, coal, (one and a half to two feet thick,) clay, sandstone. Throughout the beds overlying this coal seam, in the natural order of deposition, but, in point of fact, underlying it on account of the abnormal dip of the rocks, are many indications of lesser coal beds which as yet have not been opened. About ten feet below the nethermost sandstone (in age above it) occurs a vein of fine brick clay ten to fifteen feet thick. The limit of this stratum on its east side has not yet been reached, but the clay grows purer and more and more free from iron in that direction. On the west side of this bed the clay is not utilizable on account of the presence in it of iron, which forms a fusible double silicate and melts out, leaving the mass full of holes. A pottery has been started and bids fair to compete successfully with the best establishments of similar character elsewhere. As yet the proprietor confines himself to the manufacture of earthenware, but contemplates increasing the extent and variety of the products of this pioneer pottery, and even hopes in time to be able to rival the best English white ware.

Friday, July 16.—From Golden City to a point on the mail road nine miles east of Idaho City.—At Mount Vernon the road enters a cañon, and after cutting across a red syenite, passes into a region of finely laminated gneiss. From this point the springs become more frequent. A number of quartz veins crop out on the sides of the road. Visited a lode situated about two hundred yards south of the road and half a mile west of the stage station. The crevice was eight feet wide, and the discovery shaft ten feet deep. The quartz (which was very rotten) exhibited iron and arsenical pyrites, copper glance, and galena. The wall rock on the south side was not much weathered, whereas the proper north wall-rock had not been reached. Not far from this opening was another shaft thirty feet deep. The ore from it was rich in malachite, copper pyrites, peacock ore, and copper glance. Beautiful rhombohedra of calcite were obtained from the gangue rock. This claim was to have been sold in 1863 to parties in New York for $25,000, but owing to the effect produced by the panic among owners of Colorado mines in that year, the sale was not consummated, and the claim has lain idle ever since.

An opening on another lode still further west revealed copper pyrites, malachite, galena, and silver glance. Fine calc-spar crystals were obtained here also.

July 17.—Our route lay through Idaho City, nine miles distant. The first part of the road wound its way through masses of red and gray gneiss, intersected here and there by veins of white quartz. Now and then this gneiss alters its character, both in habitus and color. Two or three dikes of quartz porphyry cross the road.

The placer mining is carried on extensively on Clear Creek, there being sixteen sluices between the intersection of the road and creek and Idaho City. One party of the miners informed me that they averaged $12 per day per man. They had five rifles in operation.

On the banks of Clear Creek the rocks were much contorted and flexed; general dip, northwest.

The hills on the right bank of the creek are much more weathered and rounded off than those on the opposite bank. Gneiss of all kinds, heavy and thin bedded, coarse and fine grained, red and gray, with all possible combinations of these varieties, were observed. Near Idaho the gneiss becomes somewhat suddenly very micaceous.

There are six sluices in operation between Idaho and three-quarters of a mile above that city. Beyond this there is no gulch mining attempted.

At Idaho there is a hot soda spring, whose waters, however, I did not analyze. Just above Idaho is a sluice which once upon a time washed out one ounce of gold (twenty-three dollars) per hand per diem, but the best is now washed out. Near Seven-mile bridge the gneiss pitches almost vertically on the right bank of the creek, and resting upon these upturned basset edges were huge masses of gray granite.

GEORGETOWN.

The characteristic mineral of the country is zincblende, associated with galena, iron pyrites, and comparatively little copper pyrites. The most usual gangue rock is decomposed porphyry, and decomposed granite, with much quartz. The country rock is composed mainly of gneiss. In West Argentine there is considerable fluor spar occurring, as gangue rock.

Baker lode.—So far as an approximation to an average dip could be got, this appeared to be east northeast, but throughout the region the rocks are huddled together with such irregularity that nothing definite can be stated about either the general dip or the general strike of the rocks. The general strike of the veins is east of north, and their pitch nearly vertical.

Brown lode.—At most of the mines the ore is got out by hoisting, but at this one there is a tunnel driven in sixteen hundred feet above the bed of the creek to intersect the shaft. The mouth of this latter is one hundred and ten feet above the tunnel, and is met by the above-mentioned cross-cut, (one hundred and eighty feet long,) and by a drift extending (up to the date of these notes) but thirty feet out from it. The ores found in the Brown lode are native (wire) silver, antimonial silver, ($AgSb_3$,) stephanite, copper-fahlerz, polybasite, and the dark and light ruby ores. The amalgamation works below here are usually supplied with ores containing less than five per cent. of lead.

An engine of thirty horse-power drives the machinery of the mill, and in winter time warms the water intended for the wet stamps, and the building itself, by means of a steam-pipe leading to the tank containing the water. The mines are not troubled by water. In last April the miners had some trifling difficulty to contend against after the spring thaw; but this was promptly met and overcome. There are twenty stamps for wet crushing and four others each of 500 pounds weight. The ore contains about twenty per cent. lead, but this is insufficient to meet the wants of the furnace, and lead is bought to supply the deficiency. Thirty-five per cent. of lead is necessary to the carrying on of the process. There are two classes of ore which are dressed or separated by hand. The first-class ore is crushed dry and goes directly to the furnace. The second-class ore is crushed wet, and dressed by means of a circular buddle revolving from fifteen to twenty times per minute. Under one hundred ounces per ton, the ore is not treated, but is dumped out and saved in the hope that the reduced price of labor or some more economical process may enable the owners to work it to advantage.*

The ore, after having been dressed and sorted as above-mentioned, is mixed with ten per cent. lime and fifteen per cent. iron, and is subjected to a low red heat in a reverberatory furnace to reduce any argentiferous litharge that may be present. Then high heat is given, and the sulphide is converted into argentiferous lead and matt, according to the usual method.

---

* One per cent. silver equals about three hundred ounces per ton, so that one hundred ounces per ton equals one-third of one per cent.

The roasting requires from twelve to fifteen hours and the smelting twelve hours more in the reverberatory furnace. The matt, after being separated from the argentiferous lead, is stored away to be worked over at the end of every run; or if the furnace clogs up some of it is added, to clean it out by its fusibility.

A run occupies usually twenty days, more or less. In regard to the amount of work done by this company the following statement may be of interest, as giving the total from January to August of 1869.

Total number tons of ore treated in furnace, 188.

Average assay value of ore per ton, 200 ounces.

Percentage of assay value saved, 90 per cent.

The Terrible shaft is opened four hundred feet below the Brown. In April last the workmen were shut out from their shaft by the rapid invasion of water, but since then there has been no trouble. The Terrible ores have already attained a widespread reputation for richness. The main difference between them and the Brown ores is that they are richer in brittle, and the Brown in ruby, silver.*

The Baker mine.—The mill belonging to this company is situated some four miles up the cañon known as West Argentine, and on the opposite side of the creek from the Brown lode. It is one of the very finest structures of the kind ever erected in this Territory, but was not quite completed at the time of my visit. On the floor of the mill under the apertures through which the ore is to be delivered, is a drying hearth for drying the wet ores. After the moisture has been driven off the ore is crushed in Dubois's breaker and ball-crusher. The former of these machines resembles the breaker known generally in Europe under the name of "the American nut-cracker." The ball-crusher is a cylinder formed of strong iron staves, which are attached at their extremities to two stout iron disks in such a way as to leave a very small crack between each two of them. Three to four hundred pounds of iron balls are then put in with the ore and the cylinder revolved on its axis. The finely powdered rock falls through the cracks into a hopper built to receive it, and through this hopper into an iron cylinder twelve feet long and eight feet in diameter, with a helix attached to its inner surface for the purpose of continually turning the ore, and thus presenting a fresh surface for oxidation. Fire is at first applied, and this cylinder is made to rotate slowly, and in a short time the sulphur of the ore is ignited, whereupon the extraneous fire is withdrawn and the oxidation continues with the assistance of the heat from the burning sulphur. The supply of atmospheric air to the interior is regulated by means of a door to an opening at one extremity of the cylinder's axis, while the volatilized oxides of lead and zinc and silver are led through a pipe connecting with the other extremity of the axis to condensing chambers and thus saved.

After a thorough roasting the ore is let out upon cooling floors, and from that transferred to the amalgamating barrels.

The Baker ores contain much zincblende and will average perhaps sixty ounces silver per ton, though occasionally rich pockets are met with in the mine, the ores from which have given remarkably high results. Red and white varieties of fluorspar occur largely as gangue rock of the lode.

The Burleigh tunnel.—This is about half a mile distant from the Brown lode toward Georgetown. The object had in view by the proprietors of this tunnel is to intersect all the lodes whose strike is with the trend of the mountain in which it is being driven. The rock is quite hard, and only one hundred feet had been bored when it was inspected. The boring

* The average assay value of brittle silver is five thousand ounces per ton.

is done by means of steel drills worked by compressed air, the machine for driving them being mounted on a car running on rails. A steam engine outside compresses the air and forces it through pipes to the machine in the interior. It is expected that a lode will be intersected about one hundred feet further in.

The Snowdrift mine.—This mine is three quarters of a mile below the Brown lode, on the same side of the creek, and is five hundred feet higher up the mountain than the same. The ores are chiefly sulphuret, (silver glance,) and galena. Very little iron or copper pyrites or zincblende is met with. The vein is five feet in thickness, and the pay streak, (one-half of which is said to be composed of silver glance,) six inches in width. The cost of getting out five tons (including wages, &c.) was seventy dollars, and the ore averages one hundred ounces per ton.

The Griffith lode.—This lode, like the Gregory, near Central City, is the oldest as well as one of the richest in the vicinity of Georgetown. It is situated in a high hill or mountain on the right bank of Clear Creek. The shaft opening is about half way up this hill. The shaft is one hundred and twenty-seven feet deep, from which a drift has been struck fifty feet east, and ten feet west. The dip of the vein is a trifle south, though it is nearly vertical. The crevice averages perhaps four to five feet, and its north wall-rock is a syenite, while the south wall rock appears to be a weathered granite. Assays show values of from one hundred to seven hundred ounces per ton. The ore will average perhaps one hundred and fifty ounces per ton. The expectation was, when the improvements in progress had been made, to take out fifty tons of ore *per diem.* Some little trouble was experienced from water in the early spring, but not enough to hamper the efficient working of the mine.

This company owns twenty-five feet each side of the lode and three hundred *on* the lode each side of the discovery shaft.* The upper part of the north wall-rock consists of a decomposed, yellowish coarse-grained mixture of gneiss and quartz porphyry, but below it is a hard, compact syenite. The south wall-rock appears to be, above, a reddish ferruginous weathered granite, and, below, a white, compact quartz porphyry.

The following is as accurate a list as could be obtained of the principal lodes worked at the present time in the vicinity of Georgetown: Baker, (worked for three years;) Brown and Coin, Terrible, Lily, Mendota, Snowdrift, White, Elijah Hise, Wm. B. Astor, Cliff, New Boston, B. Nuckles, Belmont, Continental, Equator, Gilpin, Griffith, Comet, Magnet, Anglo-Saxon,† Young America, and Wall Street.

There are seven mills and dressing works in the vicinity.

From the Equator and Terrible the first-class ores are hand-dressed, (from the former simply broken and boxed, from the latter crushed and sacked,) and sent to the East for further treatment. The lead is not paid for. I am informed that in the New Boston mine there is in one place fifteen feet of solid galena. The same authority states that a shaft was sunk on the vein one hundred and seventy-five feet before it was discovered that the crevice, instead of five, was fifteen feet in breadth.

J. O. Stuart's mill.—This mill stands on the left bank of Clear Creek, just below Georgetown, and is built for custom ores. The greater part of the business of this mill is derived from the Equator and Terrible second-class ores. The average amount of ore put through the mill is about three tons a day, or one thousand tons a year. The process is

* See Mining Laws of Colorado.

† In the Anglo-Saxon, I am informed that native silver predominates over all other metals, but the pay streak is very narrow.

the same used in California and Nevada. Ores are never sent here for treatment which assay less than $60 per ton, and the average is about $100 per ton. These ores are roasted with salt in a reverberatory furnace and amalgamated in pans. They consist chiefly of silver glance, zinc blende, and copper (and iron) pyrites. They are first dried in an iron pan and then crushed dry in a six-stamp mill. After this they are submitted to a chloridizing roasting in a reverberatory furnace with salt. The pyrites contained in the ore is sufficient in amount to react on the cloride of sodium and set free tl e chlorine without the necessity of adding sulphate of iron, which is usually done. The material is then amalgamated in iron pans and filtered through cloths, after which it is retorted, assayed, melted, run into bricks, and stamped. The ore from the Whale lode contains about equal values of silver and gold, and will be run into bars as auriferous silver and sent East for separation.

EMPIRE CITY.

The principal mines in the neighborhood of Empire City are the Conqueror, Silver Mountain, Tenth Legion, Empire, Livingstone, Atlantic, Gold Dirt, Rosencranz, Rupp and Cross, Tom Benton, and Star, the Curtis, and Ellsworth, (the former close to Mr. Ball's mill, and the latter almost in the town,) and the Bay State. Many others look favorably, but are not mentioned, because the shafts are not yet sunk deep enough to render an intelligent opinion of their capacity possible.

The Conqueror lode.—This lode is located a mile or two above the settlement of Upper Empire. The shaft is two hundred and seventy feet deep, and the ore is all pyrites in a fine state of division. There are, as yet, no drifts commenced, but the ore is shoveled out into buckets and dumped out as a mass, resembling moist sandy clay, interspersed with fine crystals of iron pyrites. The engine, which is of twenty-two horse power, hoists out in forty seconds. They get out two cords of ore, at from eight to ten tons per cord, in a day. This Conqueror ore assays very well, but the data of its yield I am unable to find in my notes.

The Rosencranz ore resembles that of the Conqueror. The crevice of the Silver Mountain lode is five and one-half feet thick. It had lain idle for some months previous to the date of my visit, (July,) and there were ten feet of drift snow in the bottom of the shaft when I descended it. The roof and walls of the mine were covered with fine crystals of Green Vitrol.

All these lodes were recorded as striking northeast and southwest.

Mr. Ball assures me that the general character of the gangue rock in all this district is granitic.

There are nine amalgamation mills in this (the Union) district.

CENTRAL CITY.

The Gregory lode.—This crops out near the lower end of Central City, was the first discovered in Colorado, and has been worked ever since with profit, in spite of the disturbances which have checked the development of so many other mines. At present there are seventeen shafts sunk in the lode, only three of which are being worked. The first class ore of this vein is an iron pyrites in which a tolerably constant percentage of gold is found mechanically diffused, (or as some think chemically combined, with sulphur,) but at all events in a state of very fine division. The ore assays from three to six ounces gold per ton, and is sold to Professor Hill for treatment in his smelting works.

A somewhat singular phenomenon is the occurrence, at No. 4 Gregory Lode, (Bruce's claim,) of three separate veins in a breadth of fifteen feet. These veins are named the Dead Broke, Gregory, and Foote and Simmons. They are divided from each other by thin walls of country rock, in some places two inches and less in thickness, but were virtually regarded and wrought as one vein. A little higher up on Gregory hill these veins diverge in three different directions, and at a depth of two hundred and sixty feet in the Smith and Parmelee shaft the latter two are seventy feet apart.

Smith and Parmelee mine.—This claim was wrought for the first forty feet as one vein. It there divides over a mass of country rock, and, as above stated, the veins diverge continually at lower depths. At the surface in many places, the lode in which this claim is situated appears to dip with the country rock, but deeper the latter becomes almost horizontal, while the vein continues its course downward as a true fissure vein. At a depth of two hundred and sixty feet work was conducted on the north vein, and a cross-cut was run out to the Gregory lode in which there are one hundred and sixty feet of good ore which has not yet been stoped out. The level in the Gregory vein has been run east and west eighty feet.

The breadth of the vein is, on the average, two feet, and of the iron or pay streak, ten inches. The average assay value of the ore is one hundred and twenty-five dollars per ton. It is sold to Professor Hill. At a depth of four hundred and fifty feet there is another level run, and this is as deep as the Gregory vein proper has been wrought. In this level the appearance of the ore is unchanged. The mill and machinery had been overhauled and put into better condition than ever before, and the management having fallen into new hands everything seemed to be conducted with an energy and attention to details which cannot fail to make the enterprise a success. Twenty-five five-hundred-pound stamps were at work, the hoisting machinery was in good order, ventilation perfect, and the stulls in good condition. The cost of these large timbers is enormous, and out of all proportion to the other appointments of the mine. One of them, eight to ten feet long, will cost ten dollars before it is in its place.

Briggs's mine.—This claim adjoins the Smith and Parmelee, and is owned and superintended by the brothers Briggs. Everything about the mine and mill indicated that work was being conducted with intelligence and care. The condition of ladders and cribbing was good. I will venture to make one suggestion of an improvement which will apply to the majority of all the mines here, as well as to this one. In some cases, where deep shafts or other dangerous places must be passed by the miners in their passage to and from their work, a proper regard for their safety should induce the company to see to it that every accident which could endanger life is guarded against.* In some few cases this has been overlooked. The Gregory and Briggs veins, together at the surface, are fifty feet apart at a depth of four hundred and fifty feet. The distance between the wall rocks varies from four to eighteen feet. The appearance of the ore improves, the lower the vein has been followed.

At the bottom of the shaft, the Gregory vein widens out to eight feet,

* An accident has since occurred in this mine by which three men were killed.

and besides the fine look of the iron pyrites, native gold is found, in very small particles, scattered over quite an extent of the pay streak.

In the mill are fifty eight-hundred-and-eighty-pound stamps.

NEVADA CITY.

The Prize and Copeland lodes.—The town of Nevada adjoins Central City and stretches away some two miles up the gulch in which it is built. The Prize vein strikes about north 10° west, and the Copeland nearly west.

The two veins come together in the shaft at a depth of one hundred and twenty feet from the surface. The drift on the Copeland has been run seventy feet west and sixty-five feet east from the shaft. At the extremity of the western outstope the vein is ten feet in width, and the ore occurs throughout the whole of it. The ore is principally zinc-blende, and assays one hundred dollars per ton. The second-class ore averages six ounces per cord. Mine in excellent condition, and timbers good. Seventeen men and two horses are employed in and outside the mine. Back and forward stoping are being carried on at the same time from the extremities of the drift. At the bottom of the shaft the vein is six feet in thickness, and contains an eighteen-inch pay streak close to each wall. The average yield per diem is three cords (about twenty-one tons. Twenty-four stamps are run night and day.

North Star lode.—The ore from this lode contains a fahlerz which will prove very rich. The machinery and appointments of the mine are the best that I saw around Central City. The hoisting apparatus, which is provided with an automatic dumping arrangement, works beautifully. Shaft mouth, dressing works, and blacksmith shop are all under the same roof. There are eight tables for blanket tailings.

Perrin lode.—The shaft house and mill belonging to the Perrin Mining Company had just been erected under the superintendence of Mr. G. A. Bradley, but had not been running long enough to enable me to gather any reliable statistics as to the amount of work which could be done per diem.

The shaft is situated in Russell Gulch. The ores of this mine comprise copper and iron pyrites, copper glance, and fahlerz. The first-class ore averages $150 per ton, and the second-class three and one-half ounces per cord. The shaft is one hundred and forty feet deep; dip of vein, seventy-eight degrees; strike north five degrees east at the shaft mouth, but the strike varies with the distance from the shaft, and the vein appears to conform to the shape of the hill. No good hanging wall has yet been reached.

The mill owned by this company is located about a quarter of a mile from the shaft house in Russell Gulch, and is forty feet square.

There are four companies running mills in the gulch above this one, which purchase their water from the Consolidated Ditch Company. Mr. Bradley, however, has a drain to Graham Gulch, two hundred and fifty feet distant, and leads the water which he obtains from it to a tank of twelve hundred cubic feet capacity. A large cistern of five barrels capacity, attached to the rafters of the mill, keeps the stamps supplied with water, through pipes suitably attached, and derives its supply from the large tank previously mentioned.

In the event of the water supply failing, there is a second tank, of two hundred and eighty-eight cubic feet capacity, which is placed at the opposite end of the mill, and into which the water from the tail sluices runs. This tank is divided into a smaller and a larger part by a partition not quite as high as its sides, over which the water pours from the former into the latter division, thereby clearing itself. An elevator conveys it from here to the cistern. By this arrangement the same water can be used two or three times.

One engine of thirty-five horse-power drives two six-stamp and two five-stamp batteries. The stamps of the former weigh six hundred pounds each, and those of the latter four hundred and fifty pounds each.

There is a separate bin opposite each battery for sorting the custom ores. The four-hundred-and-fifty-pound stamps are intended to drop thirty-five, and the six hundred pounders twenty-five times per minute.

There are in this mill eight feet of coppers and four feet of blankets; but besides this the water runs over four and one-half feet of small blankets to the tail-sluice. Two pumps keep the water constantly supplied to the cistern. The blankets are washed, according to circumstances, every fifteen to thirty minutes.

These tailings are brought into the Bartola pans and polished by arastras, nitrate of mercury and cyanide of potassium being added in small quantities to assist the process.

From this they are brought to the dolly-tub for amalgamation. These three pans save $15 of the gold, which would otherwise run out and be thrown away, per day; and Mr. Bradley hopes to be able, by the use of three additional pans, which he contemplates adding, to pay the daily wages of the whole mill personal.

The two five-stamp batteries are always worked together, but the six-stamp batteries are provided with a clotch, by disconnecting which fastening they can be worked separately.

Cleveland mine.—Excelsior lead.

Trail Creek, a few miles from Idaho City.—I visited the mill belonging to this company for the purpose of witnessing the trial of a new two-stamp steam stamp, the invention of Mr. Wilson, of Philadelphia. Two steam cylinders are mounted on heavy framework, the piston-rods prolonged below are shod, thus forming the two stamp-rods. The weight of each stamp is 500 pounds, the impinging force of the steam 1,700 pounds; which, deducting the necessary amount for friction and other losses, leaves an available blow of over 1,700 pounds. These stamps can be run 170 to 212 per minute. This velocity was not attained during the trial, but the working was so satisfactory as to leave the impression on all who witnessed it that this kind of stamp mill, with certain modifications, bids fair to supersede all others. Great attention must, of course, be paid to the feeding, to avoid throwing upon the table imperfectly crushed quartz, because from twice to twelve times as much ore as in an ordinary mill passes in a given time under each one of these stamps.

The smelting works of Professor Hill.—These works are favorably situated on Clear Creek, half a mile below the western extremity of the town of Black Hawk. There are two reverberatory furnaces, a set of rollers for crushing, and attached to the works is an assay office for valuing the ore bought.

This ore is of all kinds and comprises the richest produced by the mines. Seven tons are matted in one day, and this matt is then sent by Professor Hill to Swansea and sold. The lump ore is roasted in heaps six to eight weeks, to get rid of the greater part of the sulphur; it is then crushed in the rolling-mill and mixed with the other ores.

The tailings, consisting mainly of pyrites, are roasted in the reverberatory furnaces.

All the ores are mixed together after roasting, in such a manner as to produce a slag of the requisite fusibility. The greater part of the zinc, lead and arsenic is volatilized, a small portion only uniting with the matt and slag. The matt contains forty per cent. copper, and is the product obtained by smelting the roasted ores. How rich this matt is in silver, and how much of it is annually shipped abroad, is known only to Professor Hill and his assistant.

It is stated that Professor Hill contemplates erecting additional works for the reduction of this matt on the ground where it is produced, and the enterprise is generally regarded with satisfaction by the mining population, among whom the belief is common that the profit which Professor Hill can realize in treating these ores ought to be sufficient to enable him to spare himself this great transportation, and at the same time stop one of the many channels through which bullion flows out of the country.

COLORADO CITY, AUGUST 9.

About three miles from Colorado City, in a ravine through which flows the Fontaine-qui-bouille, are the famous soda springs, which have been from time immemorial regarded with superstitious awe by the Indians, and which are now attracting persons from all parts of the country by their beauty and supposed medicinal virtues. Three of these springs are situated on the right bank of the creek, not more than fifteen feet from the edge, and one of them (the smallest, and that giving the strongest water) on the left bank.

The first of these which one meets in going from Colorado City bubbles up through the rock into a large basin of seven or eight feet in diameter, which it has formed partly by wearing away the sides which confine it, and partly by continual deposits of its salts. This spring is called the "Beast Spring," because it is the only one of the four conveniently accessible to large quadrupeds, which drink greedily of its waters.

The next (and largest) spring on the same side of the creek is the bathing spring, and is distant from the other but a few rods. A rude roof is erected over the spot whence it issues from the rock, and the invalids sojourning at this place (of whom there were three at the date of our visit) bathe in it night and morning. The third spring on this side is the "Iron Spring," and is situated a short distance up the stream from the last-mentioned, in a thicket, which proves from its little disturbed condition that the curative powers of the water are not held in as high estimation as are those of the other springs.

The last spring, which I have ventured to christen the "Doctor," from the strength of its water, is the smallest of all, and on the left bank of the Fontaine.

A qualitative analysis of these springs with the blow-pipe gave the results which follow. The manner of conducting the analysis was as follows:

A large iron camp kettle, of four gallons capacity, was filled with the water, and the contents evaporated to dryness. The salts deposited were then collected, and, after the water itself was tested for volatile substances, analyzed.

The Doctor.—Four gallons of the water of this spring were evaporated to dryness. The salts of the residue would weigh perhaps an ounce. The mouth of this spring is about one foot in length and eight inches in width. The water contains much carbonic acid in solution. It emerges quietly from a syenite on the left bank, and flows in a slender stream into the Fontaine. A few bubbles of gas are rising continually to the surface, but the excess of carbonic acid is not proportionately so great as in the other springs. There is a comparatively small deposit of carbonate of lime in the bed of the little canal which the water has worn away through the rock, and none in the vicinity which could be traced to the overflow of the spring. The water shows no trace of volatile substances which would escape during the evaporation, except carbonic acid. Its reaction is feebly alkaline.

The salts held in solution are as follows:

Much—$NaO . CO_2$
$KO . CO_2$
$CaO . CO_2$
$NaCl$
Al_2O_3
Trace—Fe.

The Iron Spring.—This showed the presence of that metal from which it is named by a very insignificant deposit of the familiar brown oxide in its vicinity. No volatile substances in the water. Reaction alkaline.

The salts in solution were:

* $KO . CO_2$
$NaO . CO_2$
* $LiO . {}_2CO_2$
(Probably as carbonate)—$FeO . CO_2$
$NaCl$
Al_2O_3

The amount of iron in solution in the Iron Spring was unusually small; the amount of alumina being greater and that of lime less than in "The Doctor."

The Beast Spring.—This is next to the largest. A continuous line of bubbles of carbonic acid is perpetually ascending from the bottom. The taste of the water is not so pleasant nor pungent as that of the other springs.

The analysis showed—

$NaO . CO_2$
$KO . CO_2$
$Na . Cl$
S
Al_2O_3

The *Bathing Spring* was not analyzed, but its salts cannot be very different from those of the "Beast." A noticeable feature of this latter is the small per cent. of sulphur which probably is present in soda or potash alum. The ebullition of gas in the "Bathing Spring" is enormous and keeps the water in a constant state of agitation.

This spring bursts out from a syenitic rock, but by the overflow of its waters it has covered the latter with a crust of carbonate of lime several feet in thickness and much broader than is the case at the celebrated High Rock Spring of Saratoga. It is as if a white tablecloth were laid over the rock. I have never seen so violent an escape of gas except from the Salina near Kissingen, in Bavaria.

* The potash and lithia reactions with the blowpipe are sufficiently distinguishable to enable one possessing the requisite experience to recognize them with a little trouble; but in the field, where time is short and opportunities meager, it is not always easy to do this. I venture to give them both without stating which predominates, reserving the solution of this question for the first opportunity which offers in the future.

The Indians (both those of the mountains and those of the plains) frequently visit these springs, and, riding around them upon their horses, do homage to the Great Spirit which caused them to boil forth at this place, by throwing in offerings of ear-rings, bracelets, beads, and other objects of value. A gentleman, residing here temporarily for his health, was upon one occasion alarmed at the approach of a large band of Sioux, who, he saw, were in their war-paint and on an expedition against the Utes in South Park. He secreted himself and watched them. They rode around the "Beast Spring," chanting some solemn invocation, and from time to time divesting themselves of some trinket and casting it into the bubbling water. When this was concluded they all drank of the spring, and then pursued their journey. It may be interesting to the believers in the virtues of the water to know that this same band was badly whipped by the Utes, and on its return was in too great a hurry to repeat the incantation scene. My informant took over a bushel of rings and trinkets out of the spring.

CAÑON CITY.

Seven miles up the cañon, through which runs Four-mile Creek, are four oil wells, which have been sunk by a Denver company, under the direction of Mr. James Murphy, who resides by and takes care of them.

The cañon runs through cretaceous sandstones and shales. The works are very primitive as yet. Scaffoldings surmount the mouths of two of the wells, and the oil is got out by pumping.

One of these wells is three hundred feet deep, but the oil is called by the superintendent surface oil, and he expresses confidence in reaching a much larger supply by piercing some distance down. At present he can only extract a few gallons a day. Some of the oil is stored in barrels about the premises.

An analysis made of the oil by Mr. Murphy gives—

	Per cent.
Benzine	12
Good clear burning oil	50
Nitrogenous mass, containing much paraffine and paraffine oil	25
Coke and refuse	13
Total	100

Mr. Murphy states that these oil wells have been opened six years, the half of which time he has resided on the ground, and estimates the annual production of oil at about four thousand gallons.

A quarter of a mile west of Cañon City is a soda spring of delicious water which bursts out from between No. 1 and 2 of the cretaceous. The spring is small and its strength diminished by a large acequia, the water of which leaks through an aqueduct, built to carry it around a jutting point of rock, and trickles into the spring.

The taste of the water is very agreeable, and stronger than that of any similar spring I have ever seen.

A trace of iodine was discovered in the water of this spring. The salts were—

$NaO \cdot CO_2$
$MgO \cdot CO_2$
$CaO \cdot CO_2$
Trace—Fe
Al_2O_3
Trace—I

A crust of carbonate of lime is observable everywhere in the vicinity of the spring as a porous tufa-like mass.

NEW MEXICO.—LAS VEGAS, SEPTEMBER 5.

Visited the celebrated hot springs, or "Ojos callientes." These springs make their way through metamorphic rocks on both sides of the creek, and the women of the country come to wash their clothes in them, for miles around. The temperature of the water is very high, but not being able to procure a thermometer in Las Vegas, I cannot state it with any pretense to precision. I estimate it at over 150° Fahrenheit. A five-gallon kettle of water, when evaporated, left a very slight sediment at the bottom—not a quarter of an ounce.

In solution were:

	$KO . CO_2$
	$NaO . CO_2$
	Na . Cl
Trace..............	Fe
Trace..............	S
Trace..............	Li O

SANTA FE, SEPTEMBER 9.

Visited the old placer mines, which are situated in a short chain of mountains lying thirty miles or so west from Santa Fé, and on a large grant belonging to the New Mexico Mining Company, and under the direction of Colonel Anderson, formerly of the engineer corps, United States Army.

The land owned by the New Mexico Mining Company in the San Lazaro mountains is ten miles and sixty chains square. The whole surrounding country is impregnated with gold from the mountain lodes, and gulch mining there would pay richly, were it not for the deficiency in the supply of water. The company has heretofore freely permitted the inhabitants in the grant to pan out gold for themselves, and they frequently obtain in this way six dollars per hand in one day.

The old mill which was formerly here has been replaced by a new one, now nearly completed, which is situated on the side of a hill, and by a little brook which supplies water to the boiler of the steam engine. The great difficulty which lies in the way of the successful working of these mines—a deficiency of water—Colonel Anderson hopes to be able to surmount, either by means of a ditch bringing water from the Pecus, or by sinking an artesian well. The mill contains forty 650-pound stamps, intended to drop eight inches seventy-five times per minute.

The principal mines yet opened on the property are, in the order of their importance and date, the Ortiz and Brahm.

It is highly probable that there are other veins of auriferous and argentiferous quartz on the grant, but these two being the only mines as yet opened and worked, a glance at them must suffice.

The Ortiz mine was discovered and opened by a Mexican, whose name it bears, nigh seventy years ago; but the work having been conducted in the shiftless, slovenly manner characteristic of the Mexicans, it was thought advisable by Dr. Steck, Colonel Anderson's predecessor, to sink another shaft some distance from the discovery shaft (which marks the center of the grant.) This new shaft is now two hundred feet deep, well cribbed and timbered, and supplied with the best of ladders. The country rock is a granite, and the crevice is perhaps four feet in average. An incline was begun by Dr. Steck, connecting the two shafts, and was broken through recently under direction of the captain miner, Mr. McVhee, after whom the new shaft is named.

The ore is composed of iron, copper and arsenical pyrites, galena and malachite. The pay streak is of good size, and the vein is what is called a chimney vein,—that is to say, it widens out every twenty feet or so into a chimney, which "pinches up" again a little further on.

There are quite a number of these chimneys connected together by narrow veins. This is a characteristic feature of this mine, and is considered a very favorable sign. I would especially notice here the admirable condition in which everything about the mine is kept. Though not yet extensive, the work which has been done reflects the greatest credit on the superintendent and the captain of the mine.

Altogether, about four thousand tons of ore have been taken out and piled near the mouth of the shaft, against the time when the mill shall be completed, and it can be transported thither for reduction. Its average assay value has been $26 per ton, while an ounce of gold obtained by panning has often reached $19 50.

Since Colonel Anderson's administration, $35, 000 have been expended on the mines and mill.

The transportation from the mines to the mill will cost seventeen cents per ton.

The Brahm lode.—This was discovered last April by a professional prospector of Sante Fé, employed by the company, after whom it is named. The strike of the vein is northeast, and at the surface the dip is 75°, but at a depth of thirty-eight feet it dips but 45° southeast. There are three shafts upon it, the discovery shaft being now forty feet deep.

Some fine specimens of ore carrying native gold were obtained from the extension shaft. Between the discovery and extension shafts is the air shaft, from which the richest quartz has been obtained.

Levels are being driven both ways, outwards from all these shafts.

The following may be interesting, as giving an idea of the expense of working mines on this scale in this country: Two engineers, at $90 per month; four feeders, at $3 50 per day; two amalgamators, at $5 per day; forty miners, at $2 25 per day, (Mexicans;) common laborors, at $45 per month; chief mechanic and foreman of mill, at $205 per month; one carpenter, at $5 per day; three carpenters, at $90 per month; two blacksmiths, at $110 per month; captain miner, $180 per month.

The true name of the old placer mining district is the Real Dolores. The new placer mines are situated on the north face of the Tuorto Mountains and should properly be called the Real de San Francisco. Some litigation has arisen between this company and the New Mexico Mining Company, on account of a dispute as to the boundary of the latter's part.

The nucleus of the San Lazaro Mountains is a granite, which exhibits itself in the mountain, to the north of the settlement, in high conical peaks. To the south of the settlement is a mountain composed principally of metamorphosed sandstone, which is everywhere intersected by trap dikes.

Near the mill is an igneous conglomerate. This rock consists of a matrix of calcareous matter, in which are breccia of various rocks and large rounded masses of syenite. The boulders of syenite appear to be of singularly uniform size and are strewed over the rock with remarkable symmetry and regularity.

About three miles north of this settlemeut is a high hill at the northern base of which occur several coal seams. The nearest and most recently opened is a coal of fine quality, and, like all the coal observed along the flanks of the Rocky Mountains, breaks up into small parallelopipeda or rectangular prisms.

Near the entrance to the southwest opening of this coal bed are two irregularly-shaped masses of carbonaceous clay and gypsum, which resemble, at a superficial glance, small dikes. Neither of these appear to be continued above the roof or beneath the sole of the mine, though they appear on both sides.

Another bed of coal was visited, near which was a large basaltic dike, the heat from which appears to have altered the former to a modern anthracite. This coal is harder, blacker, and more lustrous than that of other veins I have seen in the vicinity of the Rocky Mountains; nor does it exhibit that singular cleavage which characterizes these beds.

Ores were given me from the San Dia Mountains and mines which looked well, but proved by a quantitative analysis to contain very little silver. The ore was a quartz containing lumps and flakes of galena.

Colonel Anderson gave me also a fine specimen of native copper, found in the bed of the creek, at a short distance above the Real Dolores.

Quite fine-looking specular iron, hematite, and some small crystals of spathic iron ore, were seen on the North Mountain, half a mile or so from the Ortiz mine. Specimens of the former were obtained.

TAOS, SEPTEMBER 19.

Twelve miles north from Taos, in the Arroyo Hondo, is a mill erected quite recently by the "Arroyo Hondo Mining and Ditch Company," under the superintendence of Mr. Stuart, of Taos, but not yet roofed over, nor in complete running order. There are twenty 430-pound stamps constructed to drop thirty-five times per minute.

The quartz of the ore is partly a ferruginous and reddish, partly a white mixture of quartz and mica. The red variety prospects the best, ("shows the best color.") On the road from the mill to the shaft from which the company expects to derive most of its ore, is a lode which occurs in the granite and bears iron pyrites, green, and a little blue vitriol. A second opening has been made higher up the mountain into a deposit of reddish and whitish clay, which shows good color in the pan, but is too sticky to wash well in large quantities. The company is at present exceedingly puzzled to know how to treat this material, and is considering the feasibility of baking it into bricks and then running it under the stamps, which in its present condition it would only clog. In any case these gentlemen hope that by sinking deeper they will strike a true crevice and good wall-rock. "Quien sabe!"

A mile or two around the edge of the mountain is situated the principal mine of the company, which is being opened by a shaft and tunnel, the former about twenty-five feet in depth, and opening some two hundred feet above the tunnel, which latter has been driven already 180 feet and will eventually intersect it.

The dip at surface is 35°, more or less, strike about east and west. A level has been run in at the shaft mouth 65 feet, and drilling prospects well all the way in. The ore is the same as that mentioned in connection with the mill.

SAN LUIS PARK, OCTOBER 1.

In the course of a long day's march from the Sawatch to Homan's Creek, in Homan's Park, or the Rincon, we passed a region where a great number of hot springs boiled up. The first of these (and the largest) covered a space of perhaps 600 square feet, and emitted a vapor which could be seen for a long distance.

Surrounding it was a marsh or swamp in which salts from the evaporation were deposited. The temperature at the edges was perhaps 110° to 120° Fahrenheit, and bubbles of gas rise in many places to the surface, and are caused by the weight of a person walking around the edges in the immediate vicinity of the soggy soil. Specimens of the salts, as they lie loosely like an efflorescence, and also of the same material in a harder form like California marble, (only not so variegated in color,) were collected, but no opportunity offered to examine them.

The surrounding country, and our road towards Homan's Creek, is for miles covered with a white deposit called by the natives "alkali," simply. It gives the landscape the appearance of being covered with snow. This "alkali" is probably composed of nitrate of potash, sulphate of lime, and perhaps other salts in smaller quantities, but has not been yet analyzed. The same deposit has been observed in the neighborhood of St. Vrain's Creek, and in the basaltic region below Trinidad on our route down to Santa Fé.

Minerals observed in New Mexico.

MINERALS OF COMMERCIAL VALUE.

Iron pyrites, copper pyrites.—Mostly auriferous. Widely distributed in veins over the flanks of the Rocky Mountains, in New Mexico, and in numerous lesser chains of granitic and metamorphic rocks.
Malachite, green vitriol, blue vitriol.—Principally from decomposition of the above, wherever the ores have been exposed to weathering. Widely distributed in veins over the flanks of the Rocky Mountains, in New Mexico, and in numerous lesser chains of granitic and metamorphic rocks.
Zincblende.—Often argentiferous. San Dia, &c.
Galena.—Often argentiferous. Maxwell's, near Moro.
Brittle silver.—Maxwell's, near Moro.
Fahlerz.—Maxwell's, near Moro.
Specular iron ore.—Real Dolores, near Ortiz mine.
Red and brown hematite.—Widely distributed. Old Placer, &c.
Magnetic pyrites.—New Placer.
Coal.—Raton Mountains, Maxwell's, Real Dolores, &c.
Cerussite.—Maxwell's.
Anglesite.—Maxwell's.
Native gold.—Arroyo Hondo, Morena, Brahm Lode, New Placer, &c.
Native silver.—Maxwell's.
Horn silver.—Maxwell's.
Titanic iron ore.—Real Dolores.
Smithsonite.—San Dia.
Silver glance.—Morena, New and Old Placers.
Light and dark ruby silver.—Maxwell's.
Spathic and micaceous iron ores.—Real Dolores.
Turquoise, ($2Al_2O_3 . PO_5 + 5HO$.)—Cerillos, between Santa Fé and the San Lazaro Mountains.

CHARACTERISTIC MINERALS.

Quartz.—Forms gangue rock of most of the veins; common. Agate, chalcedony, and silicified wood in the bed of the Galisteo. In the granites, gneisses, &c.
Hydrated oxide of iron.—Occurs with the coal beds, and colors the rocks near exposed veins, &c.
Opal.—Galisteo beds.
Feldspar.—Everywhere among the granitic rocks. Orthoclases predominant. Oligoclases also abundant. Albite is found near Moro.
Labrador.—Basalt dikes, &c.
Hornblende.—Syenites, some basalts.
Potash and magnesian mica.—Gneisses and granites, and in the greissen found near Moro.
Leucite.—Trachytic lavas near Fort Union.

Chlorite.—Diabase, Real Dolores, San Luis.
Augite.—In the basalts and chlorites.
Calc-spar.—Very common; finely crystallized in Real Dolores.
Gypsum.—Beds near Sweetwater; also occurs with coal.
Anhydrite.
Salt.—In springs at Las Vegas and elsewhere.
Heavy spar.—As gangue rock in many veins.
Pyrope.—Fort Craig.
Chrysolite.—Fort Craig.
Obsidian.—Found near old Pecos church. Fashioned into tools, as is also chalcedony.

A fine pseudomorph of magnetic iron ore, after the cubes of iron pyrites, was picked up near Santa Fé.

The fact that I could not visit the Morena mines, which are the most important in New Mexico, and the short time given for the preparation of this report, will, I hope, excuse its incompleteness, which a more careful study of the specimens I have collected will in some measure remedy.

COLORADO TERRITORY, SOUTH PARK, OCTOBER 4.

Visited the salt springs in this park. The whole country from the hither side of the Trout Creek Pass to some distance beyond the salt works is covered with the alkali before spoken of. A small creek flows northward, and in this creek the spring from which the salt is obtained discharges its water. It is collected in a box and conducted through a small channel to the buildings. These are two in number, the one in which the kettles are placed forming a long wing at the extremity of the other. The works belong to Rawlins & Hall, and the business of salt boiling was begun by Mr. Rawlins in a small outbuilding, yet standing, in 1861.

In the long wing are one hundred and sixteen large boiling kettles, and eight iron evaporating pans.

The spring water is first run into the kettles and heated. When the water has acquired a high temperature, it is drawn off into the first of two large evaporating pans, (eleven by twenty-eight feet,) and allowed to evaporate. The sulphate of lime and other impurities are here separated from the brine, which is again drawn off into the remaining tanks. The finest grained salt is obtained from the second evaporating pan, which is eleven by nineteen feet. The six remaining pans are each five by nine feet. An analysis of the salt produced was made by A. Fennell, of Cincinnati, with this result:

	Per cent.
Na.Cl	99
$CaO . SO_3$ and other impurities	1
Total	100

The strength of the water is about one part by bulk of matter in solution in nine parts of water. (I have this on the authority of Mr. Rawlins.) The company has expended over $50,000 on the works, and expects to commence permanent running immediately. When in full operation two tons of salt can be produced daily.

Messrs. Rawlins & Hall are sinking an artesian well alongside of the long wing above referred to, by means of which they hope get a stronger brine, and also to save the expense of pumping into the kettles.

Solar evaporating vats, similar to those in use at Salina, near Syracuse, New York, are also to be constructed shortly.

The company employs from six to fourteen men. The production of a ton of salt costs the company from $15 to $20, and they sell it for from $60 to $100; the miners and smelters getting it at the former price, both because they do not require it as pure as do the ranchmen, and also because their orders are invariably larger.

REMARKS.—COLORADO.

That which has given Colorado such an unprecedented forward impetus in her internal development and growth, has undoubtedly been the discovery of gold and silver in the beds of her streams and in the recesses of her mountains. A detailed history of these discoveries would be hardly in place here, especially as this has been pleasantly outlined by Mr. Hollister in "The Mines of Colorado," but it is interesting to know that the steps toward the establishment of mills, shafts, and furnaces in the center of a but lately uncivilized country, have been the same as in California and elsewhere.

The existence of the precious metals in the mountains was not arrived at by reasoning on the similarity of the Rocky Mountains to other ore-bearing chains, nor even by concluding that if gold and silver were found in one part of their extent, they would be probably also in other parts; but the rude hunter or ruder savage chanced upon a few shining grains, which excited the curiosity and cupidity of the dwellers in the States, and first one, and then two, and then more, girded up their loins for a journey to the tempting wilderness, until the spark burst into a blaze, and hundreds of men from all classes of life were drawn together by the hope of enriching themselves with bags of gold. Many of these early gold-seekers fondly imagined that they had only to pick the gold up in the region within the shadow of the great Pike's Peak, and finding that, on the contrary, their employment was one inseparably connected with vicissitudes and uncertainties, they were discouraged and went back.

Gulch or placer mining in gold countries precedes the more regular and legitimate operations as naturally as all crude undertakings precede the improvements they suggest. The first placer mining which promised to reward the undertakers or prospectors in Colorado Territory had its origin in Cherry Creek, in a mining settlement designated Auraria, and just opposite the present city of Denver. This was in 1858.

By the laws which govern the distribution of eroded materials by the agency of water, the larger, coarser, and heavier particles are invariably found deposited nearer to, and the finer, lighter, and more impalpable wash farther from, the origin of the eroding force. Thus the drying power and heat of the sun, the oxidation of the atmosphere, and the eroding force of wind and water, tear off large and small masses of the mineral veins; gravity precipitates them, along with boulders of the country rock, into the creek and rivulet beds, and the water of these streams grinds them up as in a mortar, and finally spreads them out in beds whose distance from the point of abrasion is inversely proportional to the weight of the individual particles. In this manner fine gold may be carried to an enormous distance from its parent vein, but the farther we recede from it the finer becomes the gold and the more diffused through the silicious mass, so that the difficulty of obtaining it is increased in two ways: first, there is much less gold, and second; what there *is* is present in a much more finely divided state. To one unacquainted with the facts, this second difficulty may appear not a real one; the specific gravity of gold being the same whether the metal exists in large or small particles, must render the separation of the dust from the companion rock as easy as that of the nuggets. But experience shows that, regulate the supply of water as nicely as he may, the miner always loses a comparatively large percentage of this finely divided gold by its floating off on the surface. This will be referred to again when the effect of the supply of water on the loss of gold from the tailings is spoken of.

Where the topography of the country has been such as to cause an elbow in the stream carrying the débris, or where, from any cause, an eddy has been formed, and the diminished velocity of the water being insufficient to keep the larger rocks in motion and the coarser particles in suspension, there have been deposited at certain points little islands, as it were, of irregular but generally more or less oval shape, the gulch miner finds his richest harvest. The discovery of such deposits has often led to the erroneous belief that any part of the bed of the creek will produce equal treasure if the water be but diverted from its channel, and the construction of flumes or artificial channels in places where circumstances were not favorable to a deposition of the precious metals has, in several instances, involved the misguided projectors in useless expenditure and great waste of time and labor.

The creeks springing from that part of the range opposite and nearest to this first settlement were the first to be prospected, and, in the main, more than fulfilled the expectations which had been formed of them. The statistics in regard to gulch mining are necessarily harder to obtain than those of lode mining, for in the first place the operations are conducted by one or two men at innumerable points in various creeks and streams remote from the miners' settlements, and secondly the independent conductors of this system of mining have a natural reluctance to stating the true amount of their earnings, from the fear that other parties may be led to their vicinity and thus reduce their gains.

Statistics of that kind of placer mining which is carried on away from the beds of the streams and upon the more or less decomposed outcrop of a lode, by means of water flumed from some higher level of the creek, are easier to get at and appear to be better known. I append a few facts drawn from Mr. Hollister's book, page 66.

Zeigler, Spain & Co. ran a sluice three weeks on the Gregory, and cleaned up three thousand pennyweights, their highest day's work being $495, and their lowest $21. Sopris, Henderson & Co. took out $607 in four days. Spears & Co., two days, $853, all taken from within three feet of the surface. John H. Gregory, five days, $942; Casto, Kendall & Co., one day, $225; Defrees & Co., twelve days, one sluice, $2,080; Leper, Gridley & Co., one day, three sluices, $1,009.

At the present time there are perhaps twenty points on Clear Creek, between Idaho and Golden City, where the wheels and sluices of the gulch miners are standing, but scarcely more than one-half of them are really in operation. A few such works are to be found in all the creeks issuing from the range, but their share in the annual production of gold in the Territory is but insignificant, and their value has diminished, as is always the case with this kind of mining. While "no one has ever yet seen the lower edge of a vein," a little labor will bring one to the bottom of a placer mine, which is formed by the wash of a few fragments carried from the out-crops of the veins by the rains.

It has already been stated that the valuable ores are found in a broad belt running along the range north and south. Gold, silver, copper, lead, and zinc are found abundantly in the granitic and metamorphic rocks, which form the true back-bone of the Cordilleras, and coal in the outlying and more recent foot-hills.

There appear to be two series of veins in this great mineral belt, occurring at least along the eastern slope of the Rocky Mountains; the larger, and apparently elder, having a general north and south strike, and proving in most cases barren,* and the smaller and more recent, comprising by far the greater number of the gold and silver leads, and extending down the range as far as I had an opportunity of observing it, in New Mexico, striking generally about northeast and southwest.

It would be difficult to define sharply the direction or extent of this great mineral belt, though various writers on Colorado have indulged in fine generalizations on the subject. The fact appears to be that circumstances have been more favorable to the existence of mineral veins in some rocks than in others, and that whatever may have been the great geological causes which brought these rocks into being, where the conditions are not totally different, indications of the precious metals may be expected wherever they occur. The eastern boundary of this belt is in general terms the eastern boundary of the gneissic and granitic rocks of the Rocky Mountains, but in almost every instance where outliers of these same rocks occur, investigation has proved the existence of mineral veins: (*e. g.*, Pike's Peak, which lies 150 miles east of the main chaid of the Rocky Mountains, the San Lazaro Mountains, the Cerillos, in the valley of the Gallisteo, &c.)

The first lode was discovered in Colorado by John Gregory, May 6, 1859, on claim No. 5, of what is yet called the "Gregory lode," near Central City. The history of that discovery is very interesting, as an illustration of what energy and perseverance, guided by sound common sense, may accomplish for a man.

Gregory worked this lode at first, of course, with a sluice, and got out $972 from the disintegrated surface. The news spread rapidly, and the country was soon swarming with prospectors and miners, and many other lodes were immediately discovered and worked. This was the celebrated Pike's Peak gold fever, from which the growth of this Territory dates. In almost every case the mines passed into the hands of different parties, as the getting out and treatment of the ore became more difficult from the growing scarcity of the decomposed surface ore, until at last matters were brought to a stand-still by the resistance offered by the sulphurets associated below with the gold to the process of amalgamation then in vogue. This behavior, which would have been foreseen by more experienced miners, seems to have astonished and dispirited them, and an exodus from the region was the result, which has been repeated at various times since, whenever new obstacles were to be surmounted. But while this has retarded the unnaturally rapid development of the Territory, there is no doubt that the occurrence of these sulphurets and the working of them will, in the end, prove a blessing to Colorado, by giving employment to more persons, and thus hastening the maturity of this commonwealth.

The counties of Colorado in which as yet the principal mining operations have been conducted, are, in the order of their present importance: Gilpin, Clear Creek, Park, Summit, Lake, and Boulder.

To enumerate all the lodes which have been discovered, or even those that have been partially wrought, would be foreign to the purpose of this report, and a work of immense labor; nor would such a catalogue serve the statistician as much as might at first appear, for the object of all these incipient undertakings having been to realize the greatest pos-

---

*An exception to this general rule is found in the Hoosier lode, about forty miles north of Central City. This vein belongs to the north and south class, but is nevertheless rich and profitable.

sible amount of gold in the minimum time, and the various enterprises in any neighborhood having been conducted independently of each other, by parties whose interests were never the same and often conflicting, no pains have been taken to settle questions which did not concern the values of the ores obtained; but, on the contrary, it has not unfrequently happened that investigations of the exact positions and extent of the veins were opposed to the interests of one of the parties, which thus might be proved to be working somebody else's claim.

To explain this state of things, it will be necessary to state that by a law of Colorado (see act concerning lode claims) it is provided that—

SEC. 5. Any person or persons engaged in working a tunnel within the provisions of this act shall be entitled to 250 feet each way from said tunnel, on each lode so discovered, provided they do not interfere with any vested rights. If it shall appear that claims have been staked off and recorded prior to the record of said tunnel on the line thereof, so that the required number of feet cannot be taken near said tunnel, they may be taken upon any part thereof when the same may be found vacant, and persons working said tunnel shall have the right of way through all lodes which may lie in its course.

SEC. 7. That when two crevices are discovered at a distance from each other, and known by different names, and it shall appear that the two are one and the same lode, the persons having recorded on the first discovered lode shall be the legal owners.

SEC. 8. That to determine when the two lodes are one and the same, it shall be necessary for the person claiming that the two are the same lode to sink shafts at no greater distance than fifty feet apart, and finding a crevice in each shaft, and forming a continuous line of shafts from one lode to the other shall be conclusive evidence that the two are one and the same lode.

It will be evident from this that when two parties are working on what is suspected of being one and the same claim, those who have recorded last will be in no hurry to settle the question for the sake of the statistics, and that as it costs time and money to sink shafts fifty feet apart to well-defined walls, over a distance of three hundred feet, (the legal extent of a discovery claim in each direction from the shaft,) it is not always that these comparatively recently opened lodes are thoroughly known.

In my very restricted report of the mines of Colorado, such examples have been selected as present mining here in its best phase; or rather, of the best mines in the regions I visited, such have been selected as I could personally visit and examine. Much of interest in the details of mining here has been necessarily omitted, but I hope that what information I have been enabled to obtain in the limited time at my disposal may not be without value, though submitted without attempt at arrangement, and in the form in which the notes were taken in the field.

Many knotty questions have presented themselves to the miners and smelters, among which, perhaps, the knottiest is the dressing of the second-class ores and the proper form to which to bring the tailings before they are ready for the amalgamator or smelter. It is believed by many able miners, and the complaint is frequently made, that by the use of wet stamps and careless feeding, the mill-men waste unnecessarily a great deal of gold, and from this it is argued frequently, with less justice, that the use of wet stamps is pernicious and wasteful. This is going too far, though it is true that in the treatment of the ores around Central City and elsewhere, the greatest care and attention are absolutely necessary to prevent great needless loss. Less ore put through the mills, with correspondingly greater care in its treatment, would probably be the best remedy, and this plan would very likely produce the owners as much gold as they get at present, and leave them so much the more in the mine to work.

In conclusion I would sum up the impressions I have received from the tour as follows: That the valuable ores abound almost everywhere in the granite and gneiss of the Rocky Mountains, and the economic question is not to find the material, but the capital and labor with which to work. That the country over which these investigations were made is replete with those minerals which by their decomposition are found by experience to most enrich the soil, as it is with the before-mentioned minerals of commercial value.

That the climate is healthful and delightful, the country well supplied with water, which breaks from its rocky reservoir, with few exceptions, at distances of from ten to fifteen miles, all along the base of the mountains; the communication with the East and West is becoming daily more easy, and the savages of the plains and those whose headquarters used to be the gambling hells and drinking saloons are well nigh banished from this favored domain.

That the land is being tilled and prepared to support the large population which must soon settle here, and everything smiles on that man who brings to the country intelligence and a pair of willing hands.

What stands in the way of the country's progress are the greedy speculators who wish to use Colorado and New Mexico as mills for turning money into their pockets, regardless of the interests of the growing community. The system of grants, also, which gives to one man or one company a tract of country much larger than any one individual or small corporate body can possibly properly improve, cannot fail to exercise a baneful influence on the prosperity of such a country, by keeping back the tide of hardy and industrious settlers who would otherwise pre-empt and settle up the land. And wherever such a grant exists, a backward condition of the country may be expected. To a certain extent this disregard of the interests of these two sister Territories may be observed in the manner in which certain mines have been worked. These mines have been hacked to pieces to produce ore, and the ore has been rushed through the mill to produce gold. Nothing seemed to have a claim to the consideration of such owners but the most rapid method of *realizing*, in order that the shortest possible time might intervene before, their fortunes made, they could quit the Territory and enjoy them elsewhere. In this way, valuable mines have been ruined, and thousands of dollars of the Territory's gold thrown away. It were easier to detect this fault than to suggest the remedy; but the remedy will present itself, when Colorado and New Mexico shall be filled with citizens determined to own and occupy them, and shall have slipped entirely from the grasp of those who wish merely to hire and use them. The observation above, in regard to the remedy for the present losses in dressing tailings, has been made by several persons, and it has been added that even a smaller profit from more thoroughly and carefully worked ore would in reality pay the owners better, give a more healthy tone to mining, advance it as an art, and spare millions of dollars in the end. While the adjustment of such complicated questions as these is one which must await the lapse of time and the course of events, it would be well for interested parties to consider in what way to manage their property out here so as to assure themselves against present possible loss, and of future increase in its value. To do this without radiating prosperity on all around them, and building up the wealth and power of the country, is a problem which will tax their abilities to the utmost, however great those abilities may be.

REPORT OF CYRUS THOMAS.

AGRICULTURE OF COLORADO.

DEAR SIR: Having been directed by you, in addition to my other duties, to collect such information as I could in regard to the agriculture of those portions of Colorado and New Mexico through which your expedition should pass, I have the honor to report to you that I have performed this duty to the best of my ability and opportunities. And herewith I present a partial report of my investigations, being unable to present even a complete or full preliminary report, for want of statistics, which I cannot obtain in the field, where this is written; and, also, because I have not yet received answers to some of the most important inquiries I have sent out to some of the best informed citizens of these Territories. I hope to be able, shortly, to present you a much fuller and more satisfactory report on this very important subject. I trust that even the imperfect and partial report herewith presented will be sufficient to fully justify the interest you have taken in the development of the agricultural resources of the Great West.

Although the chief object of your expedition may be to determine the geological features of these regions, and thus increase the store of scientific facts by which the great problems of nature may be solved, yet the economic value of these investigations will be shown in the increased impetus they give to the development of the agricultural and mineral resources of the country.

Our line of travel having been along the eastern flank of the Rocky Mountain range, from north to south, my personal examinations have necessarily been confined to the margin of the arable lands of these territories. And as we were constantly moving, seldom remaining at any one point more than a day or two, I have been compelled to rely upon the statements of residents for much of my information in regard to the climate, seasons, crops, &c., &c. But I have endeavored to consult the best sources of information. Two circumstances have very much favored my efforts:

First. The proper appreciation of your efforts in this direction by the citizens, and their willingness to furnish all the information and aid in their power to facilitate the matter.

Second. The fact that the passage of your expedition through the country happened to be made during harvest, and in one of the most favorable seasons, for agriculture, that has been experienced in these Territories for many years. This enabled me to make a partial comparison of the statements received from others, in regard to the yield and quality of the crops, with my own observations on these points, thus testing the accuracy of these statements. I am glad to inform you, that so far as I have been able to make this test, it has confirmed the reports which I have received from others, showing them to be reliable.

Trusting this may prove satisfactory, I remain yours, truly,

CYRUS THOMAS.

DR. F. V. HAYDEN,
United States Geologist.

Situated between 37° and 41° of north latitude, and 102° and 109° of west longitude, Colorado Territory extends east and west about three hundred and ninety miles, and north and south about two hundred and seventy-five miles, forming a rectangle containing an area of 106,500 square miles, or 68,144,000 acres. Reaching from near the middle of the great trans-Mississippi plain up the mountain slope, it laps over the summit of the great divide and rests its western border on the Colorado basin. And including, as it does, within its bounds the great system of mountain parks, and the sources of the four great rivers, the Rio Grande del Norte, the Rio Colorado, the Arkansas, and South Platte, it has been appropriately termed "The Gem of the Mountains." And, like Switzerland in Europe, it may be said to be unique in its geographical features.

Of the large area contained within its boundary lines, about four-sevenths are embraced in the true mountain region, whose snowy summits form the watershed of the continent. The remaining three-sevenths, situated, chiefly, east of 105° of west longitude, and extending the whole length of the Territory north and south, consist, in great part, of broad plains furrowed by shallow valleys, widening and fading away as they extend eastward; and, with the exception of the parks and some valleys of the mountains, contain all the arable lands of the Territory.

But since much of this latter portion, lying along the eastern boundary, is devoid of water, excepting the few streams which traverse it, the agricultural population has, as yet, been confined within a comparatively narrow strip along the eastern slope of the mountains.

In order to obtain a more correct and minute idea of the geographical position and extent of that portion of the Territory which is susceptible of cultivation, it will be necessary to consider it in separate districts. And we are not left, in this, to mark out arbitrary lines, for nature has fixed prominent lines and permanent boundaries to each. Water is the great desideratum in the agricultural development of this country, and the method of its distribution we shall find is the true key to the agricultural system of the Territory, and its turning sheds the boundaries of the districts.

Beginning at the northern part, we find the South Platte River is the outlet for all the water of this section which flows towards the Atlantic. Moving up this stream from its point of exit, near the northeast corner of the Territory, it will be seen that after crossing the 104° of west longitude it branches rapidly into its numerous tributaries. The portion of country drained by these numerous minor streams is bounded on the west by the eastern slope of the Rocky Mountains, and on the south by a high, broken, irregular ridge called the Divide, which, starting from the base of the mountains opposite South Park, runs eastward until lost in the plains. This constitutes the northern agricultural division, which, for convenience, I shall name the Denver district.

This Divide separates the waters of South Platte from those of the Arkansas, and forms the northern boundary of the second district, which is the area lying between it and the Raton Mountains. This division, which may be appropriately named the Arkansas district, is drained by the Arkansas and its tributaries. These two districts contain most of the tillable land of the Territory lying east of the mountains.

I may as well remark here, that in my use of the terms "tillable," "arable," "susceptible of cultivation," &c., I do not intend thereby to exclude the idea of the future possibility of cultivating other sections, but simply intend to express the fact, that those sections, so termed, are now sufficiently supplied with water for farming purposes.

The third district, which is the San Luis Park, belongs both to New Mexico and Colorado, and cannot be divided into parts corresponding with the arbitrary line of division between these two Territories.

The fourth division I shall make is not a separate district, as each of the others, but includes the other parks and the small amount of arable land in the mountain valleys, which, on account of the proximity of some of them to the mining districts, become important, notwithstanding their small extent. This may be called the mountain district.

It will be seen that each of these three natural districts has its great river by which it is drained; the Denver district finding an outlet for its waters through the South Platte; the Arkansas district through the Arkansas River; the San Luis Park through the Rio Grande. And as we descend to the examination of the more minute divisions of these larger districts, we must follow the natural arrangement of streams and valleys.

THE DENVER DISTRICT.

This district is naturally divided into two sections; the first including the territory north of the South Platte and between it and the mountains; the second, the territory between the Platte and the Divide.

As the first section presents more definitely and sharply the peculiar features of this country which bear upon its agriculture than any other portion, I will give a somewhat minute description of it.

The Platte, leaving the mountains some twenty miles southwest of Denver, after bearing out a short distance on the plains, runs northeast, slightly diverging from a parallel course with the east range of mountains, for a distance of about eighty miles, where it is joined by the Cache à la Poudre, and then turns eastward; thus giving the section a triangular shape, with the north side of the Cache à la Poudre valley as its base, the mountains for one side, and the Platte the other. Its general surface is a broad plain sloping from the mountain flank eastward to the river level with valley furrows, and rounded, low ridges traversing it from west to east.

The various streams which take their rise in the mountains east of the great Divide, between the waters of the Atlantic and Pacific, run nearly an eastern course until they unite with the Platte.

The first debris, and all the heavier materials, brought down from the mountains since their upheaval, have, as a matter of course, been deposited near the base. Hence as we recede from the mountains toward the east, this local drift decreases in the size of its particles and depth of deposit. Over this is deposited the alluvial stratum forming the soil, which, close to the base of the mountain, but thinly covers the boulder drift, but increases in thickness eastward. The creeks rushing down more rapidly near the mountains, cut deeper furrows through this deposit near the base than at a distance from it. In consequence of this, the terraces or ridges, which lie between the streams, are highest above the water near the mountains, decreasing as they recede from it; that is, the distance between the water level of a stream and the top of the terrace which flanks its valley is more, half a mile from the foot of the mountain than it is ten miles from the foot. This fact in other parts of our country might have very little bearing upon agriculture, but it is a consideration of vital importance to the Colorado farmer, who must irrigate his land or receive but little return for his labor; for whenever this is the case it is evident that at some point, the water can be carried to the top of the bordering terrace.

The portion of country north of the Cache à la Poudre valley, although affording good pasturage for cattle and sheep, is not generally included in the estimate of arable land, on account of its lack of irrigating facilities. Yet the Box Elder Valley is quite fertile, and will afford room for a considerable number of good farms, and the creek, though small, is probably sufficient to irrigate the red bottom of the valley.

Commencing with the Cache à la Poudre, as the northern limit of the section, which is some seventy miles north of Denver, and proceeding south, I will describe briefly the valleys according to the streams which water them. This stream, from the point where it issues from the mountains, near Laporte, to its junction with the Platte, a distance of thirty-five miles, runs through a very pretty fertile valley, which averages, perhaps, two miles or more in width, being narrow near the mountains and expanding as it recedes from them. The bottom land of the valley is flanked on the north side by a rolling irregular ridge, and on the south side by a somewhat level terrace of moderate elevation. The stream, at Laporte, is about twenty-five yards in width, clear and rapid, affording a sufficient supply of water and ample descent for irrigating the bottoms and ridges or terraces which border it.

The next stream, going south, is the Big Thompson, which runs eastward nearly thirty miles, and also empties into the South Platte. This stream, and the valley it waters, are very similar in all respects to that of Cache à la Poudre. The third, is the Little Thompson, a tributary of the Big Thompson, but, as this creek is liable to fail in its supply of water during the summer and autumn, it cannot be relied upon for irrigation. Yet its valley affords excellent pasturage for cattle and sheep, and will furnish a most excellent range for stock when the neighboring valleys become thickly settled and fenced up. Still moving south, the next stream we cross is the St. Vrain, about equal in its volume of water to the Big Thompson. It runs through a very fertile valley of varied width, reaching the Platte at a distance of about twenty-five miles from where it leaves the mountains. The bay-like widenings of this valley afford room for extensive farms, of which the settlers are rapidly availing themselves. Left Hand Creek, a tributary of St. Vrain, affords a small valley eleven miles in length. Boulder Creek, the next in order, issues from the mountains near Boulder City, and, after running somewhat northeast for eighteen miles, unites with the St. Vrain. Some of the finest farming and grazing lands north of Denver are found along this stream. At its debouchure from the mountain gorge are gathered heavy deposits of boulders and pebbles, from which, doubtless, the creek and city have received their names. Although these deposits are but scantily covered with soil, yet the fertility seems to be but slightly impaired thereby, as is shown by the fact that here is a thrifty growth of willow and cottonwood.

The bottom of this valley, like that of St. Vrain, widens out at points to a considerable extent. Continuing our course southward with the snow-covered peaks rising above the rocky wall to our right, we next arrive at South Boulder Creek, which, leaving the mountains near Marshall's coal mine, runs a circuit of some eight miles and unites with Boulder Creek. Here, I may justly say, is found the link that unites the agriculture of the plains with the mining of the mountains, the two great interests of Colorado.

Standing on the grass-covered bluff overlooking this little limped stream, the eye, as it shoots out its glance north and east over the plains, wearies itself in attempting to mark the boundary of vision. The valleys over which we have passed in our journey southward, like dim lines, are traced across the broad meadowy expanse. Rounded ridges, level surface terraces, straight foot hills, with green swarded escarpments and isolated buttes fill up the outline. Sinking into the bluff on which we have been standing, we pass alternating strata of coal and iron ore. Here they quietly rest, rich, thick, and abundant—the fuel and the metal. The one to convert the other into instruments to till the soil, to harvest the grain, to thresh and garner it, to convert it into food, to make the highway of transportation, and carry it to the miners of the mountains and the snow-bound dwellers in the far north. Such a combination is seldom seen. And, though not directly embraced in the object of this report, yet I feel justified in alluding to it, for the reason that the opening and development of these mines are intimately connected with the agricultural development of the country. The agricultural instruments now in use are brought from the States at an expense of transportation equal to their original cost. This need not be so; Colorado has her coal, her iron, and, in part, her timber. It only needs to be developed and applied to that purpose for which nature has so bountifully provided it.

Descending from our elevated position, and continuing our course southward, after passing some minor streams, we reach Coal Creek, also a tributary of Boulder Creek. But this is not an unfailing stream, and although some farms are found along its valley, yet it cannot be depended upon for irrigating purposes. Clear Creek, which passes within four miles of Denver, gives a valley of eighteen miles before it empties into the South Platte. It is already lined with well cultivated farms and comfortable houses, from which the Denver market is in part supplied with grain, vegetables, and meat. Finally, in our course southward, we reach Bear Creek, the last of the series of these transverse streams, which, after a short run of nine or ten miles from the mountains, pours its waters into South Platte. A short distance below this we arrive at the apex of the triangle before described, which contains, including the Platte Valley, about 800,000 acres of land. Of this amount about one-third is bottom land, the remainder forming the ridges and terraces which lie between the valleys. The greater portion of this entire triangular section is susceptible of cultivation, and the remainder well adapted to grazing purposes. The bottoms along these creeks vary from half a mile to four or five miles in width, giving, perhaps, an average width of two or two and a half miles. Between these valleys are the more elevated portions, forming, sometimes, rounded ridges, at others, regular terraces or benches, or rolling, gradually descending prairies, but seldom rising into abrupt hills; the whole face of the country being richly carpeted with nutritious grasses. These ridges, which border the valleys, vary in their elevation above the water level of the creeks from a few feet, out on the plains, to forty and fifty feet near the base of the mountain, and, with few exceptions, are in reach of water sufficient for irrigation.

The valley of the South Platte is undoubtedly the most important, extensive, and fertile strip of tillable land in the northern portion of the Territory. But the descent being less in this river than in the smaller streams we have been describing, ditching, for irrigation, is more expensive. Yet it is rapidly filling up with an enterprising farming population, and is being brought under an intelligent and profitable system of cultivation.

SOUTHERN SECTION OF DENVER DISTRICT.

Passing across the Platte, going south, we enter upon a section where a considerable change of scenery is at once apparent, and where the geographical arrangement is entirely different from that we saw north of the river. There we saw a regular succession of cool, limpid streams rushing down from the Rocky Mountain gorges, furrowing their way through the plains eastward to the Platte, the great central artery of the district. Here we find an irregular arrangement of long, slender streams, arising chiefly within the space of forty miles along the northern slope of the Divide. Carrying their volumes of water down this descent, they enter upon the broad, comparatively level, and somewhat sandy plains, and receiving but few tributaries, they struggle against the rapid absorption of the porous soil, growing feebler and feebler, till finally, in the dry season, they are lost, without reaching the Platte. Plum Creek, which lies next the mountains, is perhaps the only exception. It follows, then, that the tillable part of this section is confined to the valleys along the upper portions of these streams. There is also a marked difference between the valleys of these streams and those north, in this: while the latter in most cases have bottoms of greater or less width on both sides, which are flanked by terraces with graceful, grassy escarpments, the streams south, cutting through the deep sandy deposit, generally have on one or the other side steep, bluffy banks of crumbling sand, reaching the surface of the second bottom. And even the bottoms which do border these streams very often appear to be irregular flats, scooped out of the higher land which once pressed close on the central channel. But these flats, where they can receive sufficient water, are exceedingly rich and productive, yielding some of the heaviest crops of the Territory.

In regard to the various valleys of this section, and the extent to which they can be cultivated, I can at this time give but an approximate estimate.

Beginning at the base of the mountains, and moving eastward along the northern slope of the Divide, the first stream we reach is Plum Creek, which has two branches, East Plum Creek and West Plum Creek, the one flowing from the mountains, the other from the Divide. This has a run of some twenty-five miles before reaching the Platte. It furnishes water most of this distance, and has some fine bottom lands on it, a good part of which is already under cultivation or occupied.

The next stream in our course eastward is Cherry Creek, which has quite a number of small affluents entering into it from the rounded hills on each side. From its source to its mouth is a distance of some forty-five or fifty miles, affording water for irrigation the greater part of its length, but drying up near its terminus at the city of Denver. This valley is quite fertile, and tolerably well settled at the more attractive points.

The other creeks succeed each other in the following order: Running Creek, Kiowa, Wolf, and Bijou; in regard to which I have received but little information. They generally dry up on the plains during the summer and fall, affording water for irrigation from twenty to thirty miles from their sources. Their valleys are as yet but sparsely settled. On the immediate slope of the Divide, in the bottoms which flank these streams, irrigation is generally unnecessary, as a sufficient amount of rain falls to supply the crops with the necessary amount of moisture to mature them.

SOIL.

The soil throughout this district presents great uniformity in quality, as is clearly shown by the striking similarity of the plants of its different parts. It is chiefly a light loam, in which the silicious and micaceous ingredients predominate. Yet there is a considerable difference between the two sections in one respect: while in the northern the particles are coarse and sharp, in the southern they are fine and rounded, and the arenaceous portion bears a larger ratio to the whole.

But to form a correct idea of this soil, especially in the northern section of the district, it must be remembered that the streams, in passing from the snow-clad crests of the vast range of mountains to the broad prairies of the plains, sweep over the upheaved margins of the entire geological series represented in this region. And as they rush down the mountain gorges, and along the rocky cañons, they bear away with them the debris from all the strata they touch, from the primary granite to the most recent tertiary representative, mingling it together and scattering it over the plains they cross; not only the confined streams of the present era, but all the waters which have swept the mountain slope since it was lifted up by the vast subterranean force by which they were formed. The atmospheric currents driving to and fro the lighter and dry particles on the surface, have assisted in the mingling process. This combination of the various mineral elements gives to the soil an adaptability to a large variety of plants. The predominance of silicious matter renders it peculiarly adapted to the growth of wheat and oats, and the addition of decayed vegetable materials causes it to produce heavy crops of succulent and bulbous vegetables.

It varies considerably in depth; near the foot of the mountains, where the water traveled more rapidly, it has covered the boulders and gravels with a thin crust, while farther down on the plains it grows thicker as we recede from the mountains. Although the bottoms along the creeks contain a greater proportion of decayed vegetable matter than the terraces and ridges, yet the latter are equally rich in the primary elements, and by a sufficient supply of water, will produce the cereals as heavily as the former. And, as on these terraces vegetation ripens some eight or ten days earlier than on the bottoms, they possess this advantage.

In the southern section the case is somewhat different, the Divide being largely composed of loose conglomerate of well-worn particles; when these are carried down by the more slowly running water, they become more finely comminuted and worn, forming heavier beds of sand nearer the base. In consequence of this fact the water sinks much sooner than in the northern section. This deeper deposit of sand is often very apparent along the margins of the streams where they have cut away the banks.

CLIMATE.

This strip of country lying longitudinally north and south along the east flanks of the mountain, the temperature necessarily varies as the points recede from each other. And as we descend from the higher portions along the base of the mountain to the valleys of the plains the isothermal lines bend considerably northward. But the average temperature of the northern section may be compared with that along the eastern slope of the Alleghenies, in Pennsylvania, with which it very nearly corresponds. The altitude varying from three to seven thousand feet above the level of the sea, and the line of perpetual snow not far distant, the atmosphere is salubrious and remarkably free from miasmatic vapors and impurities. And as we proceed southward, although there is a gradual increase in the average warmth, yet it is partially compensated by strong breezes which stir the air during the warmer season. In the summer the heat, it is true, is often somewhat intense, especially in the valleys where the air is partially confined. But on the higher grounds the breezes descending from the mountains render it more pleasant. The air rarified on the plains rises, while another portion, cooled by the snows of the mountains, sweeps down the slopes and brings with it a refreshing coolness.

Snow generally begins to fall in October, and ceases in April, or about the first of May, in the latitude of Denver; but, as a matter of course, beginning later and ceasing earlier in the southern districts.

Although it sometimes, though rarely, reaches a depth of twelve or fourteen inches, yet it passes off almost as rapidly as it comes, seldom remaining longer than twenty-four hours. Even in the valleys which penetrate the first range of mountains in the northern section, this is also the case. Some winters but little snow falls during the entire season. As conclusive evidence of this statement, cattle are herded out during the entire winter in all parts of the Territory, such a thing as preparation for winter-feeding being almost wholly unknown. And yet in the spring they will come out in as good order as those of the States which have been housed and fed day by day. The Mexican horses or bronchos will also winter out during the season, like the cattle.

The troublesome factor in the great problem of the development of the agricultural capacity of the vast western plains is the supply of water. Furnish this, and the fertile prairies and valleys east of the Mississippi will soon find a strong rival contending with them in the grain marts of the world for precedence. Furnish this, and the "Great American Desert" of old geographers will soon become one mighty field of flowing grain. Furnish this, and the few other minor impeding factors will soon be eliminated. The streams rushing down from the mountains slacken their course on the level plains where the great battle between moisture and aridity begins. Is there any power in the human grasp to assist nature in this struggle, and turn the scale in her favor?

Before attempting to give a direct answer to this question, I will state some facts connected with this matter, and then advance a theory, which, if correct, is of great importance in developing the agricultural capacity of this country.

When we reached the Cache à la Poudre, at Laporte, I heard it remarked that this stream now, and for a few years past, has been sending down a larger volume of water than it formerly did. I thought little of the matter at the time and let it pass, simply noting the statement. But when I reached the next stream in our journey south, the same thing was repeated in regard to other streams in that section. And to confirm the statement certain streams were pointed out, which, up to about 1862, had been in the habit of drying up annually at certain points, which since that time at these points have been constantly running. This caused me afterwards, during the whole length of our journey along the eastern flanks of the mountains, to make this a special subject of inquiry.

And somewhat to my surprise, I have found the same thing repeated at almost every point as far south as Las Vegas, in New Mexico, and no opposing testimony. Streams bearing down heavier volumes of water than formerly; others becoming constant runners which were formerly in the habit of drying up; springs bursting out at points where formerly there were none; acequias allowed to go to decay because they have not been needed, &c. Even the Arkansas, as late as 1862 and 1863, was dry, from Pawnee to the Cimarron crossing, but such a thing has not been known since. Seven or eight years ago it was not uncommon for the Pecos to dry up, but now such a thing would be looked upon as a strange event. And, in building Denver, a mistake was made in relying upon the dry bottom of Cherry Creek, which shortly afterwards sent down a rush of water to warn them of her slumbering powers. Nor does this wholly exhaust the testimony on this point, for, in addition thereto, is the uniform assertion of those who have resided in the Territory for ten or twelve years or more, that for six or seven years past there has been a gradual increase of rain. It is a common expression of the Mexicans and Indians that the Americans bring rain with them.

All this, it seems to me, must lead to the conclusion that since the Territory has begun to be settled, towns and cities built up, farms cultivated, mines opened, and roads made and traveled, there has been a gradual increase of moisture. Be the cause what it may, unless it is assumed that there is a cycle of years through which there is an increase, and that there will be a corresponding decrease, the fact must be admitted upon this accumulated testimony. I therefore give it as my firm conviction that this increase is of a permanent nature, and not periodical, and that it has commenced within eight years past, and that it is in some way connected with the settlement of the country; and that, as the population increases, the amount of moisture will increase.

It may be objected that the population bears so small a proportion to the extent of the country, that it is unreasonable to suppose it could have any influence on the climatic conditions. I admit the force of the objection, but at the same time the facts stand out too boldly and clearly to be passed over, and the coincidence is so striking, that, until the peculiar conditions surrounding the matter have been carefully studied, the objection ought not to be pressed. That there are peculiar conditions connected with the section of country under consideration, cannot be denied. Hence to know the effect the introduction of an active population will have upon the hygrometric conditions of this country, these peculiarities must be carefully studied. I believe that the great problem of settling the plains, if ever solved, must be done by commencing with the eastern slope of the Rocky Mountain range and gradually moving eastward. This is the plan which nature herself has pointed out. The perpetual snows of the great central axis are the sources of the various streams which rush down upon the margin of these plains, but chiefly sink in their effort to cross it. Let the population gather around the points where these burst from the mountains, and as it increases pushing out on the plains eastward, and I believe the supply of water will accompany it.

If this theory is correct it is worthy the attention not only of the scientist but of the citizens and authorities of the Territory, and also of the national government. A railroad line running along this eastern slope north and south would doubtless give an impetus to the settlement of this part of the Territory exceeding all that the lines crossing it at limited points (though necessary) can possibly do. It would set the great power in motion which, moving onward, would ultimately bring into use that vast body of land which by common consent has been consigned to perpetual inutility.

Such a theory may, and doubtless will by some, be considered chimerical, but before it is condemned some effort to confirm or refute the testimony given ought to be made. And I volunteer the suggestion that it would be well for the government to make a small appropriation with which the Commissioner of Agriculture could send out an agent to investigate this matter more thoroughly. Even should a more thorough examination overturn and reverse the testimony I have adduced, his labor need not be lost, as he could, while proceeding with this, gather a host of facts in regard to the agricultural capacity of our Territories, which would be of great value to the stream of emigrants pressing westward from the States.

I am aware the present season has been an extraordinarily wet one; but I have carefully endeavored to prevent its leading me astray, always extending my inquiries to a series of years, and calling attention to the unusual amount of rain this year, that it might not unduly warp the information received.

The excess of rain during this season I find has been felt most sensibly north of South Platte and between the Raton Mountains and Las Vegas. In the latter section there are some creek valleys where the rain last season was so excessive as to injure the crops, as, for instance, the Rayada. And the present year, crops in many valleys not furnishing water for irrigation have been and are maturing finely, as the beautiful one in which Fort Union is situated, which is as fresh and green as the banks of the Susquehanna.

Hail-storms are of not unfrequent occurrence during the summer, and sometimes do considerable injury to the growing crops. I have frequently, during our passage through the country, noticed fields of corn torn into shreds. But, as a matter of course, these storms are always quite limited in their extent.

POPULATION.

Colorado is pre-eminently a mining country; its mineral wealth having recently brought within its bounds most of its present population. Eagerly searching after the metallic riches which lie buried in its rugged mountains, but little attention has been paid to the cultivation of the soil. Therefore the data from which to draw conclusions, in regard to the adaptation of its soil and climate to the growth of any particular cereal or fruit, are very meager. Yet enough is known to show that, by proper cultivation, this Territory will produce as fine and as abundant crops of wheat and oats as the most favored section of the Union; and that the western border of what was once designated "The Great American Desert" will, at no very distant day, present its broad fields of golden grain. This is no wild fancy of the brain, but the inevitable result of the march of events now rapidly moving onward. That the high anticipations of the most sanguine will be fully realized I do not claim, but the derogatory reports of disappointed fortune-seekers will, ere long, be disproved by a multitude of experiments.

At this time not more than one-fifth or, as some contend, one-eight of the population of the Territory are actually engaged in agricultural pursuits, the great portion being in some way connected with the mining interests or business arising out of them. But the one must draw the other—those who mine must eat—and the heavy expense of bringing food from the States is working out its own cure. The necessity for moving forward the agricultural interests of the country are being felt and acted on. A territorial fair has been in operation for a few years, and is exciting considerable interest among all classes of citizens. Even while I am writing this portion of my report the annual fair is in progress at Denver, which, I very much regret, I have been unable to attend, but I will endeavor to ascertain all of general interest connected therewith.

CEREALS.

Of the cereals, wheat, oats, barley, and corn, grow readily and produce very good crops, when properly cultivated and irrigated.

Wheat grows well throughout the length of the Territory, north and south, and even as far south as Bernalillo, in New Mexico. So far as I have seen, and can ascertain, the following portions of these territories are the best wheat-producing sections, viz: the creek valleys north of South Platte; the South Platte and Arkansas valleys in Colorado; and in New Mexico, the Moro and Taos valleys, and the south end of San Luis Park. Besides these, there are, as a matter of course, valleys which will produce as fine wheat as those named, but these are the most extensive. The Platte Valley alone can supply, if made to yield all it is capable of yielding, the whole of Colorado with all the wheat necessary for her present population. And I am informed by Colonel Charles McClure, of Santa Fé, that the Taos Valley can be made to produce sufficient wheat to supply the entire demand of New Mexico. Until a better method of cultivation is introduced than the rude plan of the Mexican population, the capacity of the latter Territory will not be known. But, as I design considering the agriculture of the other sections of Colorado and New Mexico separately, I will confine myself to those portions of the former Territory now under consideration.

With the exception of two or three fields, spring wheat is the only kind raised. But this is not so much because winter wheat fails as it is owing to the difficulty of preparing the ground in the fall for sowing winter wheat, the ground being so dry and hard that it cannot be plowed. And if an attempt is made to soften it by irrigation, the experimenter soon learns that while one portion of his ground is scarcely moistened below the surface, the other portion is a mass of soft mud. But at any point from Clear Creek south, where sufficient rain happens to fall at the right season to moisten the ground, winter wheat sown produces a fine yield, and, as a matter of course, ripens much earlier than the spring wheat.

The usual time of sowing is March and April, though sometimes farmers, even as far north as the Platte Valley, succeed in getting their wheat in during the month of February, yet the greater portion is sown in April. Singular as it may appear, when we notice the difference in latitude between Cache à la Poudre and Santa Fé, yet it is a fact that the harvest season comes on later in the vicinity of Los Vegas, Sante Fé, Taos, and San Luis Park, than it does in the northern section of Colorado. During the present season, I see from my notes taken as we passed through the country, that wheat was cut in the vicinity of Denver between the 25th of July and 10th of August, and at Cache à la Poudre a few days later, while at Los Vegas harvest came on the latter part of August, and in the Taos Valley it was as late as the 18th of September, and in San Luis Park some wheat is yet standing, (September 23,) although the frosts set in as early as the 12th of this month. I am unable, at present, fully to account for this, but suppose it is chiefly attributable to the cold winds from the surrounding mountains and the cold nights. The average harvest time, in the sections of Colorado under consideration, may be set down about the 10th of August.

The amount grown per acre often reaches forty and fifty bushels, and there are some well-attested instances where the yield has been as much as seventy bushels. Mr. W. R. Thomas, associate editor of the Rocky Mountain News, who made, during the harvest of 1868, an extended examination of the crops in most of the valleys of eastern Colorado, estimates the average yield at twenty-eight bushels per acre. In this estimate, the absolute return when measured is the criterion, no allowance being made for bad culture, losses, &c. From a careful examination of his figures, and his method of obtaining the data upon which they are founded, as well as from the personal inquiries I have made while passing through the territory, I am satisfied he does not exceed the true average, but rather falls below it. Where proper care is given to this cereal, and it reaches maturity without serious damage from the destructive grasshopper, or other agency, a yield of thirty-five bushels per acre may be expected.

From his report for that year, which has already been forwarded to the Commissioner of Agriculture, in connection with a short report on the agriculture of the northern section of Colorado, it appears that the wheat returns from the various valleys of Eastern Colorado foot up (including estimates of the valleys omitted) nearly thirty-five thousand bushels. At an average of twenty-eight bushels, this shows that about twelve thousand five hundred acres were sown in wheat in 1868. If any statistics for 1869 are brought out by the present fair, I will try to obtain them in time to append them to this report. Most of this wheat has been grown on the creek bottoms, yet I have no doubt but that the ridges, or uplands, within reach of irrigation, will yield as heavy crops as the bottoms, and, as shown by experience as well as theory, would ripen some eight or ten days earlier.

All the varieties which have been tried appear to grow well and to bring out their several peculiarities. But those chiefly sown are the Chile, Siberian, White Sonora, Blue Stem, Canada Club, and Egyptian or Seven-head. The White Sonora is most prized on account of the beautiful white flour it makes, and its heavy yield, though it does not weigh as much to the bushel as some other varieties. In New Mexico, the Mexicans have but two varieties, the "areno blanco" or white wheat, and "areno nigra" or black wheat; the white wheat corresponding very nearly with the White Sonora. The measured bushel of Colorado wheat, if well cleaned, will weigh from sixty-two to sixty-four pounds as an average. But in comparing this wheat with that of the States, it must be remembered that the grain is perfectly dry, having been raised by irrigation, and as a matter of course having received no moisture on the ear. In this respect it corresponds with the California wheat, requiring to be moistened before grinding. Therefore a given bulk or weight of this wheat will yield more flour than the same bulk or weight of wheat from the States.

I think I am justified in saying that no part of the Union can produce better flour than Eastern Colorado, in respect to its clear, pearly whiteness, richness in gluten, and ease with which it is converted into bread; and, like the flour of the Pacific Coast, it will doubtless bear transportation to any part of the world without damage from climatic influences.

As the expedition happened to pass through the Territory during harvest, I had a very good opportunity of comparing the information I had received with the appearance of the wheat crop of the present year, from which I am satisfied the statements I have received are not exaggerated.

The wheat crop, so far as I have seen it, is very good, and I suppose will be one of the heaviest ever known in the Territory, and this is also true of the part of New Mexico through which we passed.

Although irrigation has some advantages, as that of removing fear of drought, yet it also has its disadvantages, one of which is, that it does not cause wheat and oats to ripen evenly in the same field. I have frequently noticed fields of these cereals where some spots were fully ripe, while others were yet quite green. But as the grains do not appear to shell out as easily here as in the States, this does not cause the same difficulty here as it would there.

In order to give some idea of the time of harvesting wheat throughout the section over which the expedition passed, I will give from my notes the condition of the crop at several points, with the dates at which we passed those points.

July 2. *Laporte, on Cache à la Poudre.*—Wheat in bloom. Harvesting generally takes place here about the 1st to 10th of August.

July 7 *to* 15. *Fisher's Ranch, on Clear Creek, four miles from Denver.*—Along the valley of this creek and that portion of Platte Valley in the vicinity, the farmers are cutting wheat, though the fields appear to be ripening very unevenly. Crops excellent in appearance.

August 6. *South bank of Platte, near the mouth of Plum Creek.*—Harvest just ended; standing shocks indicate a very heavy yield.

August 8. *On the north slope of the Divide near West Plum Creek.*—Harvest nearly closed; some wheat and oats yet standing.

August 9 *to* 13. *Colorado City.*—The farmers in the vicinity of this place in the midst of the wheat and oat harvest, both these cereals appearing to ripen simultaneously here.

August 16. *On the banks of Arkansas, five miles south of Cañon City.*—Wheat harvest along the banks of this stream appears to have closed at least two weeks past, the weeds almost hiding the stubble.

August 17. *A few miles west of the Arkansas River, behind the first high ridge.*—Saw the farmers cutting wheat.

August 21 *to* 23. *Trinidad.*—Wheat harvest in progress.

August 24. *Richard Wooton's, on Raton Mountains.*—Farmers cutting wheat.

August 25. *Rayada, New Mexico.*—Wheat harvest is over, having closed about a week previous to our arrival.

September 2 *to* 5. *Las Vegas.*—Wheat harvest in progress.

September 17 *and* 18. *Taos.*—Wheat harvest in progress in this valley.

September 21. *San Luis, on the Rio Culebra, Colorado.*—Wheat not all cut.

This record presents the strange fact that at the southern extremity of our route, the harvest season comes in later than at the north part of Colorado. But it should be borne in mind that this route was limited to a narrow line along the immediate base of the mountains; a similar record of a line further east might present a very different state of facts, but I have no data upon which to found a comparison.

Oats are grown with ease, and yield abundantly wherever they have been tried in the Territory; in fact there is no part of the Union where heavier crops of oats can be produced than here. Instances are reported where as high as one hundred and twenty-five bushels have been raised to the acre, but these are extreme cases. I have ascertained quite a number of instances where the yield reached from forty-eight to fifty-five bushels; and these not little garden patches which received extra care to report to fairs and societies, but crops taken from extensive farms under ordinary culture: in one instance from an aggregate of 7,000 bushels, in another 5,000 bushels, actual measurement. "Volunteer" crops will come up year after year from the seed scattered during the previous harvest.

The statistics gathered by Mr. Thomas give an average of thirty-five bushels for 1868. The aggregate amount of this cereal raised in the

Territory for that year exceeded that of wheat, but from my observations I am inclined to the opinion that for 1869 the wheat crop will be the larger of the two.

The soil and climate of Eastern Colorado seem to be well adapted to the growth of barley, which, so far as it has been tried, yields a bountiful return for the labor bestowed upon it. But the demand has not been sufficient to induce the farmers to grow it extensively.

In regard to rye my information is defective, but from all the information I could gather in regard to it, I do not think it yields as good crops as the other cereals named. I have seen but very few fields during the course of our journey this season, and even these presented an inferior appearance.

I find, since I have made a more thorough examination of the corn crops south of Platte Valley, that I was somewhat mistaken in the opinion I expressed in the report of the northern section of Colorado, forwarded through General J. A. Logan to the Commissioner of Agriculture. I there advanced the opinion that the statistics of the southern portion of the Territory would show a considerably larger yield than the northern section, but actual examination has shown me that the portion over which our expedition passed possesses few if any advantages as a corn-growing region, over the section north of the South Platte. From the Cache à la Poudre to Santa Fé I found, with very few exceptions, but one variety, the Mexican, presenting in the field a very great similarity. Although moderate crops can be produced in almost every tillable portion of Eastern Colorado, yet I must admit that it falls far behind the Mississippi Valley as a corn-producing section. Mr. Thomas gives the average yield as twenty-five bushels, and, contrary to my first impression, I now think this estimate is very near correct. The best fields I saw in the course of our journey were on the Arkansas, a short distance below Cañon City, and near a little village a short distance north of Santa Fé, named Santa Cruz. But even these I do not think would yield more than thirty or thirty-five bushels to the acre; possibly they may reach forty as the extreme. I know it is contended by some that the valley of the Platte can produce as heavy crops as the States; but if this has ever been done, the instances are rare and cannot be relied upon in fixing a general average. And this corn is, when produced, of an inferior quality.

I have no desire to underrate the agricultural capacity of the Territory in the least particular, but must state my firm convictions reached under the most favorable circumstances.

It may be that by careful experiments some varieties may be found which will prove adapted to the soil and climate, but I think there are climatic obstacles in the way of growing this cereal which cannot be overcome. But it should be remembered these remarks apply only to the sections lying along the eastern base of the Rocky Mountain Range. In this part of the Territory and in San Luis Park the nights are very cold even in the warmest part of the season, and this, doubtless, retards the growth. Besides this, the frosts set in early and prevent those varieties from maturing which require a greater length of time to complete their growth.

But, as before intimated, there are some facts connected with the maturing of cereal crops in this country which cannot be satisfactorily explained until the climatology has been more thoroughly studied. Perhaps when the botany and topography have been more thoroughly worked up, it may aid in explaining these anomalies, for such they certainly are.

VEGETABLES AND FRUITS.

Irish potatoes seem to be perfectly "at home" in Colorado, growing readily and abundantly, except in the extreme southern portion, and even here, until we pass the line into New Mexico, quite good crops are raised. Not only does this tuber grow well in the valleys east of the mountain range, but even far up in the narrow defiles nine and ten thousand feet above the level of the sea, wherever space and soil can be found, they yield quite bountiful returns to the labor bestowed upon them. The northern section, from Box Elder to the South Platte Valley, I think is decidedly the best potato region west of the Mississippi and east of California; not only in regard to the amount of the crop raised, but also in respect to the quality of the tuber.

The first new potatoes we met with large enough to eat were at South Boulder, July 6. Even at this early date in the season they were of quite good size, rich and mealy. Their rapid growth in very favorable spots sometimes causes a defect, which is also occasionally seen in the Western States—a vacant space in the center, lined with a dark internal skin.

No finer region for keeping this vegetable through the season can be found than Colorado; its pure, dry atmosphere renders it easy to keep them perfectly sound the entire year; so true is this, where proper care has been taken, that when cooked it is often impossible to distinguish the old crop from the new.

Some of the finest patches I saw growing were along the margin of Clear Creek, between Denver and Golden City, where the stream bursts its way through the lofty mesas which here flank the mountain range.

Onions grow finely, except in the extreme northern part of the Territory. The Mexican variety, which is found south of the Divide, grows to a very large size, one having been measured by Mrs Colonel Hart, at Fort Garland, whose circumference was seventeen and a half inches; very often they are found weighing two and three pounds.

As we approach New Mexico, going south, they have the finest and mildest flavor of any onions I have ever tasted, which seems to be peculiar to the climate and soil of this country, for when they are taken from here and planted in other sections, although they may grow well, they appear to lose this peculiar flavor. I was informed at Las Vegas that both seed and onions had been sent to the States, and that, in every case, such had been the result; hence, it is supposed that this delicious flavor is peculiar to this country.

Turnips and cabbages also grow quite well throughout the arable portion of the Territory, and especially in the northern section. Instances have frequently occurred where they have grown to almost fabulous size. The former grows well and produces quite abundant crops even in the little valleys far up in the mountains.

Beans are raised in considerable quantities in the southern portion of the Territory, and are much used. But they are subject to the attack of a small insect, probably a species of *Haltica*, which often does much injury to the crop. Although I did not have an opportunity of seeing this insect, yet I saw some garden patches in Santa Fé which had been literally riddled by it.

In regard to fruits, I am not able to speak positively, as sufficient time has not elapsed since attention has been turned to them to complete the experiments which are being made. But from all the indications attending these experiments there is good reason to believe from Platte Valley south all the hardier, and perhaps other fruits, may be grown successfully. Quite a number of orchards have been planted in Platte Valley and the valley of Cherry Creek, on the north side of the Divide; also, some in the valley of the Upper Arkansas. The chief trouble in the northern portion appears to be that the young trees are winter-killed. But, doubtless, this may be prevented by mulching, which does not appear to have been properly attended to. On the Arkansas, a short distance below Cañon City, I saw a young orchard, on the farm of Mr. J. T. Smith, where the trees were growing finely. Among them I saw some peach trees which have commenced bearing this season, the fourth from the seed. The apple trees are growing finely, and so far, have had no protection during the winter. There are quite a number of varieties, all appearing to be doing equally well. The pear trees, also, are in excellent condition, but the dwarfs appear to be growing the same as the standards.

Here I also saw watermelons, citrons, &c., growing to a good size.

I was informed by Mr. Smith that he had transplanted to his garden from the mountains the native currants, gooseberries, and raspberries, but that the experiment proved a failure, the bushes not growing well and bearing no fruit. These native varieties appear to be adapted only to the mountains. It is highly probable that if they were taken east and planted in the mountains the experiment would prove a success.

The wild strawberries found in the mountains of this Territory, though small, have the most delicious flavor of any that I ever tasted; they ripen about the latter part of July.

Timber for building, fencing and other purposes is a great desideratum in many portions of the country. Yet considerable quantities of pine are found on the hills which occasionally traverse the plains, and on the foot-hills which flank the mountains.

Further up in the mountains an abundance of this timber of a very good size and quality can be obtained to supply the present need. And as in these situations water-power is always close at hand, it can readily be converted into lumber by saw-mills at a comparatively moderate cost. Although there are some scrubby species of oak found in the limits of the Territory, yet in the eastern part there is none of sufficient size to be of use for domestic purposes. In our journey southward the first oak I observed was during our ascent of the northern slope of the Divide. Along the larger creeks moderate quantities of cottonwood of medium size are found. Sometimes this is seen as much as two feet through, but as a general thing it is of rather small size.

While Colorado possesses all the iron ore and coal necessary for the purposes for which these may be used, and sufficient soft timber to supply the mountain districts and sections under consideration, yet there is an entire lack of the harder wood necessary in the manufacture of agricultural and other implements. This, unless it can be grown, must always be supplied from other sections.

One advantage this Territory possesse sover many other portions of the Union is the facility with which the most excellent roads can be made. The natural soil in the streets of Denver forms a better pavement to-day than any of the artificial pavements of the cities of the Eastern States. Being a coarse silicious sand or fine gravel it forms a road equal to the best macadamized. In some of the finer sandy bottoms in the creeks in the extreme southern section sometimes the roads become heavy. Even in the mountain districts I have been astonished at the easy ascent of the passes, through the most rugged-looking ridges. Along the Union Pacific railroad as far up as Cheyenne there are but very few cuts and none of any considerable depth. And on the road from Denver back in the mountains to Georgetown there is but one steep point, and even at this, by a little trouble and moderate outlay, a good road could be made with a very moderate grade. Along the eastern base of the mountains from Cheyenne to Santa Fé there is not a difficult point to pass, the road over the Divide and Raton Mountains being no more difficult than ordinary hills in the Eastern States. The road from Santa Fé to the San Luis Valley or Park is very rough and difficult for wagons, and the same is true of the north side of the Poncho Pass, but with these two exceptions the roads to and from, as well as through the San Lüis and South Parks, are very good and easily traveled with wagons. And for the benefit of those who desire to travel over any of these routes and camp I may remark that water is to be found at suitable points throughout the entire length of the Territory. At a few points, where the road recedes from the mountains, difficulty may be experienced in obtaining wood, but such places are few, and proper precaution during the day's travel to secure a supply will prevent all difficulty.

IRRIGATION.

With a very few exceptions irrigation is necessary throughout Colorado and New Mexico. There are some points on the slopes of the Divide and in the mountain districts where the moisture afforded by rains is sufficient to supply the crops; and, as I have heretofore remarked, for the past two years, at some other points irrigation has been unnecessary, but, as a rule, it is necessary throughout Colorado, and in making up estimates of the expense of farming in this Territory this item should always be counted.

This necessity is generally classed among the drawbacks to the agriculture of these territories, but there is some doubt as to the correctness of this conclusion, for, when we take into consideration the fact that where rain is depended upon there are frequently great losses incurred because of dry seasons, the question arises, "Is the loss by drought greater or less than the cost of irrigation?" The decision of this question must decide the point as to whether irrigation is really a drawback or not. That it is inconvenient and imposes a hardship upon the farmer of limited means, at the opening or settling of his farm, is true. But when his primary ditch is completed, if properly made, he may feel himself forever secure from loss through drought.

As heretofore stated, the streams of eastern Colorado, north of the South Platte, which run from the mountains into the latter stream, have a rapid fall, varying from ten to fifty feet to the mile. Consequently it is easy to turn the water into acequias or ditches, and requires but a short run to carry it to any moderate height.

And as the terraces of this section which flank the creek bottoms seldom rise higher than fifty feet above the creek level, generally twenty-five to thirty, they can be irrigated by acequias of moderate length; in fact, I am satisfied that there is but a small quantity of land between Cache à la Poudre and South Platte which cannot be irrigated. And when these terraces have been irrigated and cultivated for a few years I feel confident that the soil on them will prove as productive as that of the bottoms.

I understand that the rule for the fall in these irrigating ditches is one-fourth of an inch to the rod, and that this carries the water over the soil with sufficient rapidity to prevent its being absorbed.

After crossing the Platte southward to the Divide and along the Platte valley, ditching is much more expensive than in the northern section, as the streams have much less fall, and the soil absorbs the water more rapidly.

Between the Divide and the Raton Mountains along the valleys of the Arkansas and its branches, the lands can be irrigated with moderate cost, though the streams are not so rapid as those in the northern section. In this part of the Territory, as well as immediately north of the Divide, the land between the streams does not assume such regular terrace forms as those north of the Platte.

Ditching at present is generally done by plowing and throwing out the dirt, except where very large ones are required. The process is also often facilitated by using a scraper. But doubtless ditching machines will soon be introduced.

The largest acequia I know of in Colorado is near Denver, on the south side of Platte River. It is several miles in length, and cost about $14,000, and irrigates quite a number of farms.

The cost, when estimated by the number of acres irrigated, is much lessened by several farmers uniting and making one large ditch sufficient to supply the farms of all entering into the combination. The smaller side ditches, which lead off to the various fields, are made with the plow, and hence the cost of these is but nominal.

There appears to have been but little improvement made in this part of agriculture for centuries past; and, in fact, it is susceptible of but little improvement. In my more extended report, which I expect to prepare during the winter, I propose to take up this subject of irrigation more fully, as it seems to hang somewhat as an incubus over western agriculture, at least in the minds of those living in sections supplied with rain. Yet when it is considered in that broad view corresponding to the vast extent of our country, it will be seen that it is one item in the series of variety necessary to the complete prosperity of the nation—one link in the grand chain necessary to render us independent as a nation.

One advantage of irrigation, which will occur to the mind of any one, is that the crop can be watered whenever it is needed. But at the same time irrigation has some disadvantages which are not apparent until shown by experience. One of these is that the crops do not receive exactly equal portions of water throughout the same field, and, consequently, do not ripen evenly. I have seen fields of wheat and oats presenting every variety of condition in the same field, from quite green to very ripe. Every farmer will at once perceive the difficulty arising from such a condition as this. It might be supposed that when the fields have been overflowed for some days with water, the land, after the water was drawn off, would assume that condition termed "baked;" but nature generally furnishes a counterpoise to all her apparent aberration; and, following this rule, she has here suited the soil to the climatic conditions, and hence this anticipated state does not follow irrigation.

I made an effort to ascertain what the average cost of ditching is to the acre, but found it next to an impossibility to do this. The difference in the nature of the ground at different points, the uncertainty in regard to the price of labor, the difference in the sizes of the ditches, would render an average, if it could be obtained, worthless.

DRAWBACKS.

As the want of water has already been somewhat fully considered, I will omit it here.

The scarcity of timber for building, fuel, fencing, and other purposes, may very properly be classed as one of the drawbacks in this Territory—one which very soon strikes the traveler passing through the eastern section. And this applies to all the sections into which I have divided the country, except the mountain district.

The amount of cottonwood and box elder found along the banks of the streams is quite small, consisting generally of very narrow fringes along the immediate borders of these streams. This is inferior timber at best, and can afford a supply but for a short time, even when we take into consideration the fact that many of the houses are built of adobes. The mountains are generally clothed with a growth of pines, but these are often of a very inferior character, especially along the eastern slope nearest the arable lands. But as we penetrate further into the mountains, these are of a better quality, and saw-mills are being erected in the interior of the mountain districts which at present are furnishing a supply of lumber at comparatively moderate prices, as water-power is easily obtained along the little creeks. But even here, notwithstanding the repeated assertions to the contrary, I do not think the supply inexhaustible. The rapid increase of the mining operations and population in the mining sections, which are in the heart of the pine regions, is rapidly consuming, for building purposes, fuel, &c., the pines around these points. And the numerous fires which occur here, and sweep up the mountain side with a wild fury, like that of a burning prairie, are destroying vast quantities of this timber. Even now we can scarcely travel a single mile along the mountain cañons where we do not see the slopes on either side marked by broad strips of burnt timber, which appear as somber spots on the otherwise beautiful scenery.

I have no doubt but that this view will be controverted; yet when we look at the broad expanse of untimbered lands which sweeps out eastward from the mountain base, and compare it with the timbered strip in reach, it is scarcely possible to arrive at any other conclusion. But, as I am now pressed for time, I will reserve the discussion of this subject for the more extended report I desire to present on the agriculture of Colorado and New Mexico. Even now, sawed lumber has frequently to be hauled for seventy-five to one hundred miles, and even further; but the building of railroads will greatly reduce the expense of transportation.

I bring this matter forward thus prominently in order, if possible, to impress upon the farmers and citizens of this Territory the great necessity of commencing at an early day the work of planting trees. It is certainly one of deep interest to them, and every effort should be made to induce not only farmers, but all who have lands and lots, to commence this important work. I am sorry to say that throughout our journey I saw but few houses surrounded by growing trees.

In regard to the supply of fuel, the difficulty of supplying this want will doubtless be met when the various coal mines are opened, and railroads traverse the country. But the supply of fencing material, at a reasonable rate, is not so easily met. To avoid expense on the larger farms, that portion intended for cultivation is generally inclosed under one fence, and corrals are made for the stock which is in use. And I have noticed some cases where two or three small farmers have combined and inclosed their farms under one fence. Around Denver wire fences are being introduced, and will probably prove cheaper than any other kind, unless hedges can be made. If this latter kind can be made, I think they will prove the best that can be adopted, not only for the same reasons urged in their favor in the prairie States, but also because they will assist in increasing the amount of moisture, and in drawing birds, thereby tending to decrease the grasshopper pest.

Another serious drawback to the agriculture of Colorado, as well as other portions of the great trans-Mississippi plains, is the destruction of crops by the migrating grasshoppers. During some years, in different localities, these insects have proved very destructive, sometimes sweeping away in a few days the result of the hard labor of the farmer during an entire season. Yet I find, after a somewhat thorough examination, that in this Territory although occasionally very injurious, yet they are by no means so destructive as has been represented. And, as has been the case this season, the papers of this western country often imprudently spread false alarms. This arises from a neglect to distinguish the larvæ of the migratory species from those that are merely local. I am satisfied that there is but one migratory species—the *Caloptenum spretum*—which appears here in any considerable numbers. The *Oedipoda coralipes* (Hald.) is found at certain points in limited numbers, but I do not think it ever proves destructive east of the mountains.

I have noticed during our trip the former species at various points as follows:

On our arrival at St. Joseph, Missouri, June 17, we found them very abundant in the complete state, so much so that the lower parts of the walls of the hotel at which we stopped were literally covered black with them, and the hogs, which seemed to have learned the art of catching them, were enjoying a rich feast. I understood they had been moving for something over a week previous to the date of our arrival.

At Omaha, Nebraska, the 18th of the same month, I saw none of them. It is true I did not go out of the city to examine, yet I think if they had been present in any considerable numbers I should have seen them.

During our stay at Cheyenne, (from 21st to 28th June) I noticed them in considerable numbers, but in the larva state and scarcely half grown.

At Box Elder Creek, and Laporte, on the Cache à la Poudre, I saw none, although I made diligent search for them; but when we arrived at Big Thompson, two days after, (July 3,) I found them quite abundant in the perfect state. From here to Clear Creek, Denver, but few were seen. At the latter place (July 7 to 14) I observed them in moderate numbers, just entering the perfect state, but the local species were rather numerous, both in individuals and species. From here we passed westward into the Middle Park (from July 15 to 27) and all along the mountain valleys, after entering the first range, and in the park we found them in abundance in the perfect state, often rising, when the wind was prevailing, in large swarms and floating before the wind like huge flakes of snow. I traced them even up into the very midst of the eternal snows, gathering specimens from the cold surface; and, strange as it may seem, even above the snow, on the naked summits of the peaks, I saw the larvæ of this species hopping about almost as lively as those on the plains.

After crossing the South Platte, going south, although individuals were occasionally seen, yet at no place during the remainder of the journey were they seen in abundance.

One conclusion to be drawn from the foregoing facts is that, even within the limits of the eastern portion of the Territory, there are distinct local broods; for while they were abundant and active, in the perfect state, at Big Thompson, July 3, yet, on Clear Creek, about a week later, they were undergoing their last moulting, and between these points scarcely any were found.

Another important conclusion which I think we may draw from these facts is, that the mountain cañons and valleys are the primary hives from which these vandal hordes issue upon their destructive mission—important because it renders the problem of counteracting them more difficult of solution. In many other sections where this pest was met at the introduction of population, the opening of farms, and bringing the soil under cultivation, &c., has gradually brought the destructive species down to their normal condition. But here, as these mountain cañons and slopes cannot be brought under cultivation, this counteracting influence can never be brought to bear upon them. Yet, even in this case, nature has not left this evil wholly without a counterbalancing opposite. While she has made the mountain valleys and sides the hive from which her destroyers swarm, she has hid within the bowels of these lofty ranges rich mines of gold and silver, to attract thither an active and energetic population. Through these the homes of this insect will be disturbed, and the primitive broods gradually destroyed. Hence while the mountains send down the evil, they contain the remedy. And like the little wave made by the pebble dropped in the lake, which swells in proportion as it recedes until it dashes against the shore, so it is with each counteracting effect within these mountain sections; it will be felt in increasing proportion along the whole line of their migration.

I have been unable to ascertain with any degree of certainty the distance they move in one season. I am aware calculations have been made on this point from data obtained on the eastern side of the plains. What reliance is to be placed upon these I do not know.

STOCK RAISING.

I cannot at this time enter upon the consideration of this very important branch of agriculture, for the reason that I have not as yet obtained all the data necessary, and also, because I prefer to defer it until I present a report on the agriculture of New Mexico.

But I may now state generally that these Territories possess as fine grazing lands as any to be found in the west. And although much stock is raised here, yet the amount falls far behind what it should. Many who are rushing back and forth from point to point along the Rocky Mountain range, seeking for rich lodes, would probably find much more gold if they would turn their attention to stock-raising. Not only do the plains afford good pasturage, but grass of most excellent quality clothes many of the mountain slopes, and carpets the lofty mesa surfaces and elevated mountain valleys. On the top of the Divide, there is one of the most beautiful little grassy plains I ever saw, where a large herd of cattle or sheep could find rich pasturage.

The finest butter and milk I ever tasted was obtained in South Park. So delicious was the milk that the members of our party could scarcely satisfy themselves with it.

There are abundant openings for industrious and energetic stock-raisers to make money following their occupation in this country. And for the benefit of such as feel an interest in this matter, I herewith give a synopsis of the laws of Colorado, respecting non-resident stock-owners Revised Statutes of Colorado, chapter 70.

Section 1. Non-residents may herd stock in this Territory for one year by payment of fifty cents for each animal so herded, in lieu of all other taxes; on sheep, twenty cents.

Section 2. Non-residents desiring to herd cattle in the Territory, must file with the recorder of the county a certificate of the number and description of such cattle in the following form:

TERRITORY OF COLORADO, ——— *County*, *ss:*

"The undersigned, owner *(or agent of the owner, as the case may be)* of the following described animals, proposes to keep and herd the same for grazing purposes within the county aforesaid, to wit: *(describing the number of animals of each kind, respectively, with brands, if any,)* from the —— day of ———, A. D. 18—, until the —— day of ————, A. D. 18—."

Section 5 prescribes the penalty for herding without filing such certificate; which is two dollars for each head of stock, except sheep, and one dollar for each sheep.

Section 9. Non-residents driving stock from one county to another not to incur an additional tax.

* Section 10 forbids the importation of Texas cattle.

These are the principal sections which relate to the herding of stock in the Territory.

The following is a synopsis of the laws of the Territory relating to irrigation. Revised Statutes, chapter 45.

Section 1. Claim owners on the bank, margin, or neighborhood of any stream, entitled to use the water for irrigation.

Section 2. The right of way through claims of adjoining owners for the purpose of conveying water allowed.

Section 3. Extent of the right of way extends only to ditch, dike, or cutting, sufficient for the purpose required.

Section 4. Where the water is not sufficient to supply all, the probate judge to appoint commissioners to apportion it.

Section 5. If the right of way is refused by owner of lands through which the ditch runs, it may be condemned.

Section 6. Persons in the neighborhood of a stream may erect wheel or other machine for raising water; right of way therefor may be obtained.

Section 7. Ditch owners required to preserve the banks of their ditches so as not to flood or injure others.

GENERAL REMARKS.

Although unable to attend the territorial fair held at Denver this year, September 22 to 26, yet, since my return, I learn that it was well attended, and that the show of stock, farm products, and minerals was the largest ever presented at any fair held in the Territory, and the interest taken greater than any previous season. I cannot attempt to give a list of articles and premiums, but may be excused for stating that the premium on turnips was awarded to W. H. Berry, esq., of Fairplay. I mention this because these were raised on the highest part of the surface of South Park, some ten thousand feet above the level of the sea, almost at the margin of eternal snows. I have procured specimens of these, which are very large, though inferior to those that received the premium.

The crop of 1869 is larger than that of any preceding year, and is estimated at the following figures: Wheat, 675,000 bushels; corn, 600,000 bushels; oats and barley, (nine-tenths oats,) 550,000 bushels; potatoes and other vegetables, 350,000 bushels. Which, with the hay and dairy product, will have a market value of not less than three and one-half millions of dollars.

In conclusion, we may confidently assert that Colorado, at no very distant day, is destined to be one of the chief agricultural sections in the Rocky Mountain regions, yea, we may say the most important. The

* This section is by no means strictly observed or inforced.

mining regions affording a home market, it possesses a completeness within itself not found in any other section of the Union, while New Mexico will be the great central fruit and wine region.

I have on hand a large amount of notes and items in regard to the agriculture of the other sections of Colorado and also of New Mexico, which I hope to present at an early day.

o

www.ingramcontent.com/pod-product-compliance
Lightning Source LLC
LaVergne TN
LVHW021406110826
845150LV00007B/1805

* 9 7 8 1 4 2 5 5 1 2 0 8 8 *